AGUSTÍN ESMARO GUEVARA RUIZ

Generation of electrical energy by gravitational effect

AGUSTÍN ESMARO GUEVARA RUIZ

Generation of electrical energy by gravitational effect

Example of application of science: A patented experimental study

ScienciaScripts

Imprint

Cover image: www.ingimage.com

This book is a translation from the original published under ISBN 978-613-9-43793-1.

Publisher:
Sciencia Scripts
is a trademark of
Dodo Books Indian Ocean Ltd. and OmniScriptum S.R.L publishing group

120 High Road, East Finchley, London, N2 9ED, United Kingdom
Str. Armeneasca 28/1, office 1, Chisinau MD-2012, Republic of Moldova, Europe
Printed at: see last page
ISBN: 978-620-8-24624-2

Contents

Bibliographical references

Guevara, A. (2020). *La teleologla de los experimentos cientificos: el caso de la calda Ubre de los cuerpos.* [Tesis de pregrado, Universidad Nacional Mayor de San Marcos, Facultad de Letras y Ciencias Humanas, Escuela Profesional de Filosofia], Repositorio institucional Cybertesis UNMSM.

SUMMARY

The aim of this thesis research was to show that the experiment of the free fall of bodies is susceptible to change its purpose. With this research we demonstrate that the experiment of the free fall of bodies can be repeated, but with different objectives than those that motivated Galileo Galilei to perform an experiment using the inclined plane.

The hypothesis that guided our research was: The experiment of the free fall of bodies is susceptible to change of purpose. In order to confirm this hypothesis, it was necessary to invent a technological prototype that works at the moment when a suspended body falls towards the centre of the Earth; in the case of the experimental processes that allowed us to confirm our hypothesis, the body to be attracted was the mass of water contained in one or more containers.

After having confronted the theories with the results of the experimental processes, the hypothesis was confirmed, as expressed in the main conclusion of the experiment: By using a sash as a means of transmitting gravitational potential energy, with a flow rate of 10, 757 L/min, with a height of 8,56m, we manage to obtain 11, 56 volts DC.

The results of this research have made possible the granting of a patent - by Indecopi - to the Universidad Nacional Mayor de San Marcos.

INTRODUCTION

The present thesis research report offers a conceptual content abstracted from two well-differentiated aspects of human knowledge: on the one hand, this research presents purely theoretical contents based on sources that provide us with descriptive and theoretical data, basically. The other academic side is based on original data taken from scientific-experimental processes and the elaboration of a mechanical system as a material element necessary to obtain the data, especially quantitative data, obtained from the various steps and experimental processes.

As an opening to the research, theoretically, we make historical and thematic reference to the contribution that Galileo Galilei made to science and, with it, to humanity. From the intellectual contribution of the Italian scientist, we have taken some contents, which are related to some parameters of the scientific-experimental process on which Galileo relied to obtain the necessary data for the elaboration of the law of the free fall of bodies.

Our research is based on having repeated - in some way - the experiment about the fall of bodies using, for this purpose, technical means such as a belt, pulleys, shafts and containers mainly.

As far as the invention of the technological system is concerned, it was designed in such a way that when an object falls in the direction of the Earth's centre of gravity, it generates a mechanical propulsion; in this case the objects to fall were the various containers that are strategically attached to the propulsion belt, as we will see later on.

In this sense, the objective we sought to achieve at the end of this thesis research consisted in demonstrating that a scientific experiment is susceptible to change its purpose. In effect, the experiment that we took into account to exemplify the proposed objective was that of the free fall of bodies. After having carried out several repetitions of the experiment, our research objective was achieved, since we used the mass of water contained in containers as bodies subject to terrestrial attraction, and these containers, being adhered to the belt, when falling following a vertical direction, generated a certain propulsion and, with this, it was possible to generate electric energy (See figure N° 2).

In the scientific aspect, we indicate the role played by the terrestrial force of gravity in its condition of constant, in accordance with the independent variables (volume and height). It should be pointed out that we are not looking for quantitative data derived from the various scientific-experimental steps and processes to find the scientific law that would explain the phenomenon of the fall of bodies, as Galileo Galilei did, but, in our case, we are looking for the fact that, following the constant of gravity and some intervening variables in the Galilean experiment, we could obtain a product of practical utility that would benefit society: electrical energy.

As far as mechanics is concerned, Galileo Galilei made use, above all, of instruments such as the inclined plane and spheres; in fact, the aim of the father of modern science was to find the law which would explain the phenomenon of the free fall of bodies. On our part, the aim was different, as different were the mechanical instruments used; that is to say: our experimental aim was to find the way to produce electric energy; for this purpose we did not rely on the inclined plane and spheres, but on a belt, pulleys and vessels, mainly; as detailed in chapter IV.

In both the scientific and technological aspects, the formula describing the earth's gravitational force as a constant was taken into account.

With some technical improvements with respect to the first designs of the prototypes, the mechanical system invented, in its final design, was presented to the Indecopi to follow an evaluation and administrative process to obtain the patent; the same that has been granted to us, by the Indecopi, on 05 November 2020.

In order to present the steps, methods and conclusions of this research in the most detailed way possible, we have divided this thesis research schematically into four chapters:

In the initial part of the first chapter we present a brief historical reference to the interaction between human beings and the phenomenon of terrestrial attraction. In another part of the same chapter, we present an analysis of the role that philosophy, as an all-encompassing discipline, plays or should play in the face of the challenges and problems that the human species is currently facing, particularly the problems derived from climate change. In this chapter we also argue the reason for having elaborated a thesis that is based on original data derived from a series of experimental processes, conducted by someone who was academically trained in a professional academic school that is traditionally detached from experimental scientific work. At the end of this chapter we present the general objective of the research and the respective hypothesis.

Meanwhile, in chapter II, we expose the complementary character that exists between science and technology, so we give credibility to specialists who argue that technology is an application of science, taking as an illustrative example the hydroelectric power plants, which are electromechanical systems, which, in order to function, need scientific and technological knowledge, and whose product is electricity, which is a product of great importance for the progress and welfare of different societies in today's world. And, given that knowledge is transmitted by means of a language, in this chapter we also address the issue of the type of language by means of which technology communicates the knowledge related to it.

Chapter III. In this section, we present the most relevant scientific contributions of Galileo Galilei himself, his scientific production that gives epistemological support to the law of the free fall of bodies.

Finally, in chapter IV, we deal with the issues concerning our experimental processes, both in the scientific and technological aspects. Here we present the main conclusions of our experiment, the main one being the following: We managed to produce electrical energy with the following parameters and variables: By using a sash as a means of transmission of gravitational potential energy, with a flow rate of 10, 757 L/min, at a height of 8.56 m, we have managed to obtain 11, 56 volts DC.

CHAPTER I

HISTORICAL AND EPISTEMOLOGICAL ASPECTS

1.1 Summary

General objective of the thesis research: To demonstrate that the experiment of the free fall of bodies is susceptible to change of purpose.

Hypothesis: The experiment of the free fall of bodies is susceptible to change of purpose.

Analysis and content: One of the academic tasks necessary to understand the origin, evolution and development of this or that discipline of human knowledge is precisely to know its historical development. In the case of the academic formation in philosophy, understood as an academic-professional career, throughout the formative process, in more than one subject, thematic contents related to the history of science, technology and technology are taught; in addition to this, in the academic-formative process, courses have been taught that have allowed us to elaborate evaluative judgements about science and technology, we refer specifically to the courses related to epistemology.

Based on the knowledge acquired during the academic training in philosophy, we have thought it convenient to develop an investigation in which we explain and apply one of the scientific experiments developed by Galileo Galilei: the free fall of bodies.

Now, if we make a brief analysis of the social and academic context in which Galileo developed his different scientific activities, regarding the free fall of bodies, we find that -among other objectives- he pursued one objective: to find the truth about the phenomenon of the fall of bodies. After developing a rigorous experimental-scientific process, he managed to give a satisfactory explanation, since it was a scientific explanation, which was made following a scientific-experimental method, whose results he presented them expressed in mathematical language, we refer, specifically, to the law known in the scientific and academic ambit, as the law of the free fall of bodies, which in its algebraic structure includes the force of terrestrial gravity as a constant.

Centuries have passed since the aforementioned Italian researcher gave a scientific explanation to a historical phenomenon peculiar to the planet we live in; however, despite the passing of the centuries, at present, the quantitative value of the law of the free fall of bodies, expressed in mathematical terms, is being applied in different areas of human knowledge, especially in the field of technology.

In this thesis research, we explain the experimental process of an additional application to the existing applications; to make the application concrete, we methodically chose to apply the methods and principles of the experiment of the free fall of bodies in order to obtain, as the main result, the generation of electrical energy, in which process it was necessary to use the fall of certain volumes of water contained in containers of strategic size and shape.

After having carried out the experiment based on the principles of the law of the free fall of bodies, we obtained the expected result: we were able to generate electrical energy by using a mechanical system, which, in the process of its operation, includes the law of the free fall of bodies as a constant.

When carrying out the different measurements and processing of the data collected from the experimental process, quantitatively, a differential advantage of the experiment object of this research was observed, with respect to the other processes of obtaining electrical energy from the use of the gravitational potential energy of water, since in the experiment we did not use the force obtained from the hydraulic pressure, but specifically, we chose to take advantage of the terrestrial gravity force, since this attracts all the bodies located on the terrestrial surface or close to it, being one of these bodies the mass of the water.

In order to exploit the force of gravity as effectively as possible, it was necessary to invent a system, mainly of a mechanical nature; with this invention, it was possible to exploit the earth's gravity by using the mass of the water and the height as independent variables in the scientific-experimental process.

The results obtained from the experiment have allowed us to confirm the hypothesis of the experimental research, that is to say, we have demonstrated that, by means of certain methodical processes, it is possible to generate electrical energy, taking the law of the free fall of bodies as a constant. The fact that we have confirmed the hypothesis means, therefore, that we have reached an important technical, technological and scientific achievement.

From a philosophical perspective, the fact of having taken cognitive sources linked to philosophy to develop an investigation such as the one we have just described, allows us to demonstrate that philosophy has sufficient means and academic instruments to contribute to part of the solution of certain problems that are presented to us in the current scientific, social and academic circumstances; in particular, with part of the solution to one of the problems of a universal nature: The lack of energy, which is related to the types of sources of energy, since it is known that many times, the sources and processes of obtaining energy, are not the most appropriate, from the ecological point of view, since in many cases they lead to environmental pollution, since the use of these sources, generally become causal agents of the increase of the temperature on the earth's surface.

Finally, we believe that it is pertinent to emphasize that, an academic work as the one we have been explaining, is, in some way, a continuation of the scientific-philosophical task that one day, the first philosopher of humanity started: Thales of Miletus in ancient Greece, who demonstrated, experimentally, certain characteristics and properties of the physical phenomenon that makes possible the action at a distance, this demonstration was based on the observation of the attraction of small particles, after the Greek philosopher made a rubbing of objects that met particular characteristics and properties (amber). What the aforementioned philosopher did, in our opinion, was the first experiment on phenomena related to electricity and electromagnetism; moreover, "it is from the word 'electron' (amber: yellow, in Greek) that our word 'electricity' has been formed" (Garcia, 1957, p. 14).

1.2 Motivation

The Peruvian geography, by its very nature, presents very divergent characteristics, such divergence of characteristics can be established from different points of view; for example, the Peruvian territory has from snow-capped mountains to deserts; it has forests on the coast, in the highlands and in the jungle, but these differ greatly from each other. Another indicator of geographical divergence is the fact that the

coastal plain and the Amazonian plain are separated by the Andes Mountains.
If we consider the division of Peruvian territory into Coast, Highlands and Jungle, we find that in different areas and at different times, there were civilisations that developed thanks to the fact that they were able to face the natural adversities offered by the areas in which these civilisations settled; but that, in turn, they were able to take advantage of the advantages offered by the places in which these diverse social groups settled.
Archaeology, as a science that studies the vestiges of the diverse cultural manifestations of the past, has demonstrated that, in these three regions of Peru, there was the presence of pre-Inca and Inca cultures, cultures which are conventionally called pre-Hispanic cultures. However, it is important to clarify that, although these cultures had a type of knowledge that allowed their preservation in time and space, this knowledge did not meet the conditions and characteristics necessary to be qualified as scientific or technological knowledge. Therefore, following the criteria established by many historians, archaeologists, philosophers, epistemologists, among other specialists, we will say that the knowledge possessed by the pre-Hispanic cultures was a technical or pre-scientific knowledge. Precisely the vestiges of the cultural manifestations of the past provide tangible evidence of this type of knowledge.
It should be noted that one of the characteristics of science is its artificiality and its degree of abstraction, characteristics that we do not find in the archaeological remains left by pre-Hispanic cultures.
At this point it is pertinent to point out that when human beings have had limits to their natural abilities, they have sought and - in some cases - succeeded in "prolonging" their natural abilities by relying on technology, technology or science; Thus, for example, as an example of a combination of technique and practical mathematics, in order not to forget a large number of elements, the human being relied on objects which together represented a certain number, then, when counting it again when necessary, even if time had passed and the number had been high, the individual recovered the numerical data without major difficulties; in this case, he was prolonging one of his natural capacities destined to store data: his memory. This is an example of precientffic artificiality which was certainly practised by the different civilisations in the ancient world.
Already in modern times, to cite two illustrative examples, we find the telescope and the microscope as instruments that make possible an extension of the natural human ability to see.
Alphabetic writing was and is, without a doubt, a form of prolonging the memory of the human being, and it is the one in which human artificiality is most evident throughout the process of the evolution of the species *Homo sapiens sapiens*. Writing, as an artificial memory, not only stored data referring to mathematical knowledge, but also data expressed grammatically, including explanations, descriptions, definitions, etc.
This type of writing, as a whole, made it possible to form a kind of "mega-memory", and even today we make use of certain data stored on physical supports that were written down thousands of years ago. And, today, every day, the human mind continues to accumulate an infinity of data on physical and digital supports, by means

of writing, in all parts of the world, data to which the human memories of the future will have access without major efforts, and this event will be repeated every time the human individuals of the future deem it convenient.

Writing, besides being understood as an artificial memory, is also an artificial means of transmitting knowledge, since it was and is through writing that knowledge was and is communicated, particularly knowledge derived from scientific work, especially the results of a research process.

This type of artificial memory was not the recipient of the knowledge available to pre-Hispanic cultures, therefore, pre-Hispanic knowledge did not advance beyond technical knowledge, at the practical level, and mystical explanations of some natural phenomena, at the abstract level.

But it is pertinent to point out that the lack of alphabetic writing was not an obstacle for *Homo sapiens sapiens* belonging to pre-Hispanic societies to develop portentous works that, even today, are the object of scientific study in many cases.

As in different spaces and prehistoric and historical times, in the world, the force of earthly gravity was also a natural phenomenon with which the men of the pre-Hispanic cultures interacted; some of the evidences that give support to what we have just affirmed are found, for example, in the construction of bridges, construction of great fortresses, construction of numerous terraces, hydraulic canals, roads, etc.In all these works, which were not built 'overnight', the humans of this part of the world and of this period of time, surely, sometimes had gravity as a factor adverse to their interests, but, at other times, they had it as a factor favourable to their practical objectives.

In order for our statements just made to be confronted with facts, more precisely with facts of a historical nature, we have decided to verify *in situ* one of the great and portentous works of high "hydraulic engineering" developed in part of our Peruvian soil and in pre-Inca times; we are referring to the amazing Cumbemayo canal, located in the department of Cajamarca.

Indeed, Cumbemayo is an archaeological site located about 19 km southwest of the Peruvian city of Cajamarca, at an altitude of about 3,500 metres above sea level [1].

[1] The Archaeological Park and Intangible Area, recognised by Law 24047, are located on a plateau of treeless grasslands, among the hills carved by wind erosion. Like a stone serpent, the Cumbemayo channel rises at the top of the Pacific-Atlantic divide, and follows a path that is sometimes straight, as if it wanted to pierce the clouds, and sometimes zigzagging. Cutting through the living rock, and heated by the sun's rays, it carries the crystal-clear water for nine kilometres to the farmland. Its origin is not known, but

Source: *Own elaboration.*

Everything indicates that more than three thousand five hundred years ago, even without ceramics, it was the scene of complex invocation rites, the same ones that Andean anthropology cannot yet decode. Its original traces have no other motive. There is no other way to explain the thousands of hours invested in carving 853 metres of living rock to transfer one litre of water per second. It could well have been built in another way. If it had really been made with a utilitarian purpose, that is to say, for irrigation, it could have had larger dimensions, but the initial dimensions, in the living rock, of 0.35 to 0.50 m wide and 0.10 to 0.30 m deep, only obey one objective: to be part of a stage for the magic of the rite. If we add to this the fact that every certain stretch of the cave there are striking petroglyphs that, like the writings of who knows what forgotten prayers, signalled the process, we must conclude that this was the reason for its construction. We do not know how the men of Cumbemayo communicated with their gods, or with the God of Water in particular, but the study of the geometry and symbols of their petroglyphs is exciting and surprising. Two kilometres of slope for every metre of length is enough to prove the mastery of agronomy that our ancestors possessed (Deza, 2012, pp. 13 - 14).

Photo N° 2: *Cumbemayo Canal, straight line under a bridge.*

Source: *Own elaboration.*

Photo N° 3: *Cumbemayo Canal, straight line.*

Source: *Own elaboration.*

Source: *Ancajima (2013, p. 06).*

About this type of vestiges linked to pre-Hispanic hydraulics in our country, different studies have been made, such as the one by the anthropologist and physician John Earls, who mentions that:

Inca irrigation of the highlands made extensive use of the supercritical flows in the canals [...], the irrigation of the highlands was basically for risk control, while in the drier places near the coast there could not have been any agriculture without irrigation (2006, p. 119).

To prevent disasters that could have been caused by the runoff of waters that descended attracted by the pull of the earth's gravitational force, pre-Hispanic settlers built canals: they were fighting against gravity; instead, by using the waters that

reached the coastal valleys attracted by the force of gravity, they were harnessing the force of earth's attraction for a collective, hence social, benefit.
The mere fact of having built an irrigation canal on a slope, through which water flows from a high part to a low part, allows us to understand that in this process three natural elements were present: the force of gravity, a height and the volume of water, and as an element produced by the human mind there was the practical knowledge reflected in the construction of some kind of canal.
With the arrival of the Spaniards came a wide range of expressions of European culture, both at the technological, technological and scientific levels. Among the novelties coming from the core of European culture, which had a significant impact on the ancestral cultures of the Tahuantinsuyo, were, in our opinion, religion, the Spanish language, currency, writing and the wheel.
There are a series of examples of a technical and technological nature that confirm that part of the European knowledge found, in Peruvian territory, suitable material means in which it could be applied and materialised; thus, a clear example of the aforementioned application was the mills designed to grind grains and cereals that worked with the force produced by falling water.
In this type of application of European knowledge in the territory of the South American continent, we find a clear example of complementarity between European knowledge and the material qualities that the territory of this part of the world possesses; thus, the Peruvian soil offered the stone to design and carve the wheel; At the same time, the Peruvian geography provided the necessary slopes for the masses of water to move, attracted by the force of gravity; nevertheless, this example of the application of European knowledge in Peruvian territory is a type of knowledge at a pre-technological level, as is demonstrated by the remains of a hydraulic mill which we have documented by means of photographs taken *in situ* .[2]

[2] The interaction of *Homo sapiens sapiens* with the force of earth's attraction has evolved over time, i.e. it has progressed from the simple use of water by means of canals, through the construction of windmills, to hydroelectric power stations. In order to provide tangible data of this evolution, at this point of the research we present the vestiges that show the interaction of the human being with the earth's gravity, in this case, to take advantage of it, but only at a pre-technological level. These are vestiges of mills whose propulsion comes from the fall of a certain volume of water, but which today no longer operate this type of mechanical machines, as they have been replaced by mills for grinding grain that operate with electric motors. These vestiges, which bear witness to the use of the force of earth's gravity, are located in a rural area of the department of Cajamarca - Peru.

Photo N° 5: *Panoramic view of the remains of a grain mill, based on the force of falling water.*

Source: *Own elaboration.*

Photo N° 6: *Stone wheels, as a remnant of a water-powered mill. Interiors of photo N° 5.*

Source: *Own elaboration.*

Over the centuries, human knowledge has evolved; however, in certain cases, even today we find products in which the complementarity of European knowledge and the characteristics and properties of the Peruvian territory can be observed, but the result of this complementarity is already at a technological level, and therefore, a result of the application of science.

Moreover, at present there are several examples of hydraulic technologies installed on Peruvian soil that work thanks to the existence of natural elements, such as a height, a volume of water in interaction with the earth's gravity, and there is also a slope through which a certain volume of water flows, and, as elements produced by human creativity, we have grammatical and mathematical writing, and also a whole electromechanical system that allows the potential energy of water to be converted into kinetic energy, which in turn is converted into electrical energy: this is an example of human knowledge reflected in science and technology called a hydroelectric power station.

Indeed, the construction of a hydroelectric power station is a pathetic example of the fact that the Peruvian geography has offered and still offers adequate and appropriate characteristics in which scientific knowledge can be applied and, thus, some kind of technology can be developed. It is, moreover, an evident example that, although the ancient Peruvians used these natural elements (volume of water, height, terrestrial gravity) at a technical and/or scientific level, nowadays, these same natural elements and some others can fit into the game of scientific and technological development that human knowledge makes possible today.

Undoubtedly, if the pre-Hispanic settlers did not manage to combine these natural elements with the intellectual elements of scientific and technological character, it

was because they were not within their reach or were not available; however, in present times, in our opinion, it is the task of the researchers of this sector of knowledge to do what they were unable to do: to complement the scientific and technological knowledge with certain geographic characteristics that the Peruvian territory offers; these characteristics, in many cases turn out to be very favourable to apply science and technology in order to reach the welfare of some social sectors of today.
To cite some examples of complementarity between nature and knowledge: the Peruvian territory offers the winds to generate wind energy; the deserts to generate photovoltaic and wind energy; and the large waterfalls on both slopes of the Andes to generate hydroenergy, which, by the way, the hydroenergy potential is not yet exploited to one hundred percent of its natural potential.
It is true that the "flag" raised by western knowledge was writing, and the writing par excellence suitable for science has been and is mathematics; for its part, the Peruvian territory has certain characteristics that allow it to perform as a receiver of western knowledge, when it comes to applying science, in the specific, scientific and technological knowledge; Therefore, the task of the Peruvian researcher -among other aspects- should consist of analysing and calculating where and how the orographic characteristics of the relief of the Peruvian territory can be successfully complemented with scientific and technological knowledge from the West and from other parts of the world.
In the case of our technological innovation, we have managed to establish a successful complementarity between the slope (characteristic of the Andes Mountains), the height, the volume of water, the force of earth's gravity and the scientific law established by the European Galileo Galilei, the law that explains the phenomenon of the free fall of bodies; to achieve the aforementioned complementarity we have relied on electromechanical devices of a technological nature, obtaining as a result a technological system that allows the generation of electrical energy and, with this, to cover one of the priority needs of the rural populations of the Peruvian Sierra; when speaking of need, in this case, we refer, specifically, to the lack or scarcity of electrical energy as an indispensable resource in the current state of civilisation.

1.3 Statement of the research problem

It is true that, since prehistoric times, the human being started to use his first tools in order to satisfy, mainly, some of his vital needs. From that time to the present day, every day man has carried out many of his activities in order to achieve his well-being; it should be pointed out that a significant number of these activities were related, in some way, to the force of the earth's gravity; that is to say, humans have always developed activities, either to face the force of gravity, or to take some kind of advantage of it. To mention a very everyday example: in the very fact that a human individual takes a step, it is true that: at the beginning of the step, the individual "faces" the force of gravity y lifts his foot; however, as soon as the foot begins to descend until the end of the step, the force of earthly gravity, turns out to be beneficial; and, this was a phenomenon that occurred in prehistory, in history and, without doubt, it happens nowadays.
Continuing with the historical perspective, it can be understood that man's

relationship with the force of gravity became a dynamic interaction, which has been repeated, in different forms and in different spaces, from those archaic times to the present day; however, despite the fact that the force of gravity has played a significant role in the evolution of history, for a large sector of the world's population, this important influence of the force of gravity on various activities of daily life of all mankind goes unnoticed at the present time.
The history of science shows us that different human groups, with different degrees of clarity, identified and distinguished the force of terrestrial attraction; since then, trying to live in harmony with it, humans have always sought to invent technological and/or technological instruments, either to face the adversities offered by terrestrial gravity, or to take advantage of it; these inventions have been nothing more than exclusive products of reason, the same ones that are currently reflected in science, technology and technology (examples: the equation describing the free fall of bodies, a hydroelectric power station, the lever, etc.)
Following and overcoming various stages of human evolution, the possibility arose of overcoming knowledge of a technical and utilitarian nature, thus passing to stages in which, gradually, philosophical, scientific and technological knowledge began to predominate, respectively. It was in these stages, in which reason began to predominate, that people emerged who, exceptionally, wondered about the reason for the presence of the force of gravity on the earth's surface.
According to various historical sources, there were many people who tried to give a rational explanation to the phenomenon of the fall of the different bodies.
One of the first intellectuals of the antiquity that tried to explain the referred phenomenon, was the philosopher Aristotle, in his work titled: *Physics,* in which he explained the distinction of the types of movements, so much to terrestrial level, as well as in the ambit of the other planets and of the stars, also.
But centuries later, the person who asked himself why bodies fall and who, at the same time, gave a satisfactory answer, because he had a scientific character, was undoubtedly Galileo Galilei, and he did so by relying on the experimental scientific method.
At the time when Galileo developed his experiment aimed at explaining the fall of bodies, there was a gap in scientific knowledge about the truth about the natural phenomenon referred to terrestrial attraction; it was in these circumstances that the Italian scientist proposed and then succeeded in explaining the phenomenon of the fall of bodies by relying on the method of experimental scientific character and, whose results he presented them expressed - preferably - in mathematical language.
In this historical context, the search for a reasonable explanation of why bodies fall was a scientific necessity, i.e. there was a need to know the truth about the earth's gravity as a natural phenomenon.
In fact, Galileo Galilei achieved his goal and, as a result, he left us as a valuable scientific legacy the law of the free fall of bodies, derived from experimental data, since the law of the free fall of bodies is the law of the free fall of bodies.

Galileo made use of a method of time measurement based on a water clock. He marked the position of the dial on the inclined plane at equal intervals of time. From these marks, Galileo realised that the distances travelled during the time intervals kept an odd-numbered proportion: 1, 3, 5, 7. Since the proportions held with more inclined planes, the same effect had to occur in the free fall. The time taken to traverse each unit of space is 1,3, 5, 7..., which means that it takes one unit of time to traverse the first section; at the end of the second section it has taken a total of 1+3 = 4 units of time (Corcho, 2012, p. 111).

From these preliminary data, the law is deduced in an algebraic expression:

$$g = 9.8\ m/s^2$$

As can be seen, this law is expressed in mathematical language, which has an operational character, since, in the formal structure of the referred concept, more than one variable and the gravity constant (*g*) concur in a single formula, but which, when operating mathematically, the referred concept explains the truth about a single phenomenon: the force of terrestrial gravity.

The mathematical concept in reference, Galileo obtained it following the hypothetical deductive method, for which it was necessary to follow an experimental process and, to communicate his results to the world, he used algebra, a mathematical discipline that - by the way - was a valuable legacy of the Arabs, which is why the mathematical concept that explains the phenomenon of the free fall of bodies, has the form of an equation, therefore, in its formal structure there appear certain variables and constants ordered according to the formal rules of algebra.

Galileo's equations of motion are used to know the position and velocity of a body throughout its motion in a vacuum and can be used with great precision in a gravitational field, i.e. when dropping a body from a certain height. (Fernandez, 2012 pp. 66 - 67).

At present we have sufficient knowledge about the fall of bodies, expressed in mathematical concepts, and this explanation is epistemologically called a scientific law; this law, for the academic and scientific community, constitutes a valuable intellectual resource, which has allowed and allows, among other things, the development of various types of technologies and, with it, the improvement of the living conditions of many sectors of world society.

In present times, whenever we believe that it is convenient to repeat the experiment of the free fall of bodies, in different contexts, we do it, generally, for didactic purposes, since the natural phenomenon already has a scientific explanation, so that, however many times it is repeated, the results that are arrived at will always be the same. An emblematic and historical repetition of Galileo's experiment is the experiment that,

In 1971, Apollo 15 astronaut David Scott dropped a feather and a hammer on the lunar surface, to verify that both reached the ground at the same time, given the absence of atmosphere on our satellite and, therefore, the lack of friction, so that Galileo's equations of motion could be fulfilled. "Which proves that Mr. Galileo's ideas were correct", commented Scott at the end of the famous experiment, as a tribute to the Tuscan (Fernandez, 2012, p. 67).

This emblematic repetition of the Galilean experiment is a historical landmark that supports our hypothesis of the present investigation.

As researchers in the field of epistemological and scientific knowledge, we clearly understand that the phenomenon of the fall of bodies already has a satisfactory explanation; but, in our opinion, what still remains to be done is to increase the number of applications of the Galilean law for the benefit of mankind; These applications necessarily require innovations at the scientific and technological level, which shows that the application of the law of the free fall of bodies is a task left by Galileo, but which is not yet completed, although at some point in history it was repeated on lunar soil, even. Therefore, one of the duties of today's managers of

science and technology is to create means and conditions in which this valuable scientific instrument can continue to be applied as a valid means to solve certain social problems that afflict mankind today.
In an attempt to fulfil -in some way- part of the task, we have invented a mechanical system that works, among other things, thanks to the force of the Earth's gravity, and whose observed efficiency in its operation is directly proportional to the problems of shortage or lack of electricity in certain parts of the world.
It should be noted that, to repeat the experiment in order to find an algebraic equation that allows to explain the phenomenon of the free fall of bodies, at this stage of time, would be unnecessary, and we have clearly distinguished this unnecessaryness, which is why, in our experiment, the main objective was no longer to find the truth about a physical phenomenon, but, rather, our objective was to reach the efficacy of the law of the free fall of bodies, efficacy that is susceptible of being explained by the law of the free fall of bodies, The main aim of our experiment was no longer to find the truth about a physical phenomenon, but rather to reach the efficiency of the law of the free fall of bodies, an efficiency which can be evaluated and qualified with percentage concepts at the moment when the invented technological system is put into operation; Therefore, this truth about terrestrial gravity - which is explained by means of an algebraic expression - increases its radius of action, because, beyond the purely scientific scope, it goes beyond and reaches the sphere of social reality, too.
The aforementioned increase in the radius of the reach of truth, which is explained by means of a theoretical instrument called the law of the free fall of bodies, reaffirms or reinforces the true character of the Galilean algebraic formula, since, maintaining unaltered its purely scientific and mathematical essence, it is also susceptible of being applied to the solution of social problems; this application makes evident the degree of efficiency and the practical utility of the aforementioned algebraic formula.
Of course, the application of the law to various practical cases is tangible proof that the formula is in a position to play very important roles, even beyond the strictly scientific field, to reach the highest possible level of efficiency, which will have a positive impact on the field of social reality, which is characterised by infinite needs.
Now, in the process of our thesis research, our hypothesis is as follows:
The experiment of the free fall of bodies is susceptible to change of purpose.
In this respect, it is important to clarify that this statement is not limited to being a mere proposition expressing a hypothetical affirmation; but, on the contrary, in our condition of cultivators of the scientific-experimental method, although we have repeated, in some way, the experiment of the free fall of bodies, the difference lies in the fact that we have not sought to find a scientific law that explains a natural regularity, but rather, following a different path (technological path), we have sought - from the repetition of the Galilean experiment - to solve a problem of the free fall of bodies, we have sought - from the repetition of the Galilean experiment - to solve a problem of a social nature, since, from the use of scientific-experimental processes, from the use of the law of the fall of bodies, plus the invention of a suitable technological system, we have succeeded in producing electrical energy by means of the use of the gravitational potential energy of water.
It is opportune to emphasize that the process of production of electricity by means of

our invention, has turned out to be more effective in relation to the systems of production of hydroenergy produced by conventional technological systems, since, by means of the invented technological instrument we have been able to take advantage of the gravitational potential energy of the water, in such a way that, we have been able to convert part of the force of the terrestrial gravity in electrical energy, thus achieving the objective of our experimental research, which consisted not in discovering a law to explain the phenomenon of the free fall of bodies, but in demonstrating that the application of the law of the free fall of bodies, as a medium and scientific instrument, can effectively contribute to the solution of a problem of a social nature: the shortage of electrical energy, which, in turn, correlates with a problem of an ecological nature, which affects the whole world and requires urgent solution: climate change.

1.4 Aim of the research

To show that the experiment of the free fall of bodies is susceptible to change of purpose.

1.5 Justification

Humanity, in almost all ages of prehistory and history, has faced problems of different kinds, some more serious than others; others, more or less universal; in the face of these problems, human beings have reacted by offering different types of answers, according to the knowledge and capacity of explanation that the species *Homo sapiens sapiens* has had in the face of any problem. However, it must be pointed out that, although in their origin societies explained many problems taking as a support the elaboration of myths, with the passing of time, rational explanations gradually emerged, which, in the beginning were given by philosophy alone, and later, scientific explanations were added to philosophical explanations.

Certainly, the present age is not exempt from problems that have a high degree of seriousness; therefore, philosophy is called upon to respond - in some way - from its own domain of knowledge. We consider that one of the problems on which philosophy must act, or at least express its point of view, is climate change. Since philosophy is characterised by the fact that it is a discipline that cuts across all the sciences, it must therefore try to find some kind of solution to the problem we have just mentioned.

There is no doubt that climate change today is more than a mere problem, but a threat of considerable magnitude, not only for the life of the human species, but also for all biotic beings living on planet Earth. In this respect, as early as 2001, Bunge warned:

> The problem, then, boils down to this: if we are to overcome the global, planetary crisis of our time, we need more scientific research and more technology than ever before [...]. All that is known is that either we face the crisis rationally and realistically, or our civilisation, or even our species, will become extinct. The great dilemma of our time is rationality and realism or extinction (2001, p. 150).

Faced with such a problem of considerable magnitude in the world, professionals from different areas of human knowledge have the duty to contribute to the solution of this problem, since we are convinced that "climate change in the world involves many different fields: geology, climatology, oceanography, physics, chemistry, ecology, biology, astronomy, etc." (Earls, 2007, p. 17); particularly, those of us who have chosen to study philosophy at a university are also called upon to provide, at least, an alternative solution, which means a solution, if not a total one, at least a partial one.); particularly, those of us who have chosen to study philosophy at a university are also called upon to provide, at least, an alternative solution, which means a solution, if not total, at least in part, to the problem mentioned above; in this way we will avoid being part of those philosophers who, according to Bunge, "are not aware of what is discussed in other departments or of what happens in the society that houses and nourishes them. They only read philosophical literature, and write exclusively for colleagues" (2001, p. 104); on the point of view of the author just cited, it is worth noting Bacon's position on the inseparable link between philosophy and practical reality: "When philosophy is cut off from its roots in experience, where it sprouted and grew, it becomes dead" (Bacon, quoted by Corcho, 2012, p. 22).

That is why, as researchers, based on scientific and philosophical principles - as an attempt - we have taken the method and the law of the free fall of bodies as a means to find some kind of solution to a social problem. For this law to be applied with the highest level of efficiency in technology, it was necessary to invent a technological instrument, in such a way that, by combining the scientific law, the experimental process and the invention of the mechanical instrument, it was demonstrated that it is possible to take advantage of the force of terrestrial gravity in interaction with the mass of water, to obtain, as an expected result, the generation of electrical energy from a constant force existing in terrestrial space and time (terrestrial gravity) and a renewable source of energy (water). Having proved that it is possible to generate electrical energy, taking the force of gravity and the mass of water as factors, confirms our research hypothesis.

In this way, we have shown that certain actions can be taken from a philosophical point of view to contribute to solving the problem of climate change. The mass production of the electromechanical system that generates electrical energy undoubtedly contributes to minimise the negative environmental impact and, at the same time, to protect the environment without depriving the access and use of the comforts and conveniences that human civilisation has today; therefore, our contribution from the field of philosophy agrees with the State and Government policies assumed by many countries in the world aimed at combating the also called: climate change; including, at present, the vast majority of countries in the world seek to address the problem, from the implementation of an economic-political system of international cooperation and multisectoral; that was how a multinational body emerged called: The Conference of the Parties (COP) which is the supreme decision-making body of the United Nations Framework Convention on Climate Change (UNFCCC, for its acronym in English); it should be noted that Peru is also a member of that international body.

The first COP was held in Berlin in 1995. To date, 25 COPs have been held,

> At its twenty-fifth meeting organised and chaired by Chile, which took place between 2 and 13 December in Madrid, the 197 Parties that make up the treaty -196 nations plus the European Union-, will seek to advance towards the implementation of the agreements that have been determined in the Convention that establishes specific obligations of all Parties to combat climate change (COP25, 2019, n. p.).

This multi-sectoral and international body refers to its vision:

> The whole world is in a process of transformation towards truly sustainable development. Raising ambition with a balance between mitigation and adaptation is key. For this we need the participation of

states, local governments and the private sector. The COP must favour concrete climate action, ensuring an inclusive process for all parties and the formal integration of the scientific world and the private sector. Our challenge is to achieve a transition towards scaled-up action that is perceived by the public. Climate change is a reality today, not in 50 years' time (COP25, 2019, n. p.).

As we can see, the formation of the Conference of the Parties is basically a political response; that is, it is a response agreed upon by the political representatives of various countries around the world.

Now, being faced with these circumstances, related to climate change, the question arises: ^What is the response that we, as professionals in philosophy, should offer?

It is well known that philosophy is characterised, among other things, by giving an answer to certain types of problems. In our case, our answer would be to have used a kind of scientific experiment as a means to generate electricity for the benefit of society, without the production of such energy contributing to the increase of the greenhouse effect in the world.

As a guarantee of the effectiveness of the operation of our technological contribution, our mechanical invention, once completed, has been subjected to various technical and scientific evaluations. Specifically, the invention object of the present thesis research, after having passed different experimental stages, at the time passed the internal evaluation by the General Direction of Research and Technology Transfer, attached to the Vice-rectorate of Research and Postgraduate of the Universidad Nacional Mayor de San Marcos; then, the invention also passed the evaluation carried out by the

National Institute for the Defence of Competition and Protection of Intellectual Property (Indecopi)

As stipulated in the internal rules of the University and, in order to obtain the patent for the Universidad Nacional Mayor de San Marcos, we assigned the intellectual property rights to the University, before a notary public (See ANNEX N° 2), this in accordance with the Intellectual Property Regulations of the Universidad Nacional Mayor de San Marcos, which in its Article 15 referred to the ownership of patents of invention and utility models; in its paragraph "a", refers to that:

The UNMSM has the ownership of the research carried out by teachers, researchers, students, thesis students, administrative staff when they are developed as a result of the exercise of the functions inherent to the contractual employment relationship or in the case that the invention results from specific agreements for scientific research and the development of science, technology and innovation in which the UNMSM is involved (2018, p. 10).

By virtue of the above, the University, before the Indecopi, presented the application to proceed with the registration and subsequent obtaining of the corresponding patent (See ANNEX N° 3); the mentioned patent application was admitted for processing, satisfactorily and, and we even had the opportunity to participate in the XVII national contest of inventions and industrial designs, organised by the same Indecopi (see ANNEX N° 4). After having passed the different stages of evaluation and qualification in accordance with the rules laid down by the World Intellectual Property Organization (WIPO), Indecopi granted us the patent on 05 November 2020.

On the basis of the achievements and achievements obtained in the administrative-legal field and with the results obtained in the technological aspect, we are in the conditions to affirm that the hypothesis that guides the present investigation has been confirmed; and moreover, the confirmation of the hypothesis is supported by a factual fact, which, in itself, means an achievement, both at administrative level, as well as technical, technological and scientific, we refer to the obtaining of a patent for the Universidad Nacional Mayor de San Marcos, as a result of a thesis research (see APPENDIX N° 7).

Now, seen from a philosophical approach, all this research work allows us to demonstrate the role that philosophy plays and/or should play in the face of the challenges that humanity is currently facing. In fact, from our perspective, we are convinced that - as it did in the past - philosophy can contribute to the solution of global problems, taking as a means of support the various forms of knowledge that together are part of the intellectual heritage of humanity.

Finally, we believe it is worth emphasising that research of a scientific-experimental nature such as that described in the present thesis report is, in a way, a continuation of the scientific task initiated by the Greek philosopher Thales of Miletus; by the way, we believe it is worth emphasising that research of a scientific-experimental nature such as that described in the present thesis report is, in a way, a continuation of the scientific task initiated by the Greek philosopher Thales of Miletus,

What makes the name of this philosopher more frequently remembered today is that he is considered to be the first to have spoken of magnetism and electricity. He knew, in fact, that amber has the property of attracting light bodies after having been rubbed, and that a magnet stone attracts iron (Garcia, 1957, p. 14).

1.6 Research hypothesis

1.6.1 Grammatical structure of the hypothesis

The experiment of the free fall of bodies is susceptible to change of purpose.

1.6.2 Logical structure of the hypothesis

If an electro-mechanical system includes the law of the free fall of bodies as a constant in its operation, then **the experiment of the free fall of bodies has changed its purpose**.

The expression of the hypothesis by means of the formal language of logic indicates that we are dealing with a conditional molecular proposition. In fact, as it is an expression that forms part of the informative function of language, the statement is a proposition, therefore, it is susceptible of being qualified as true or false. And, since it is a proposition of molecular type, and because of its type of operator it is a conditional, we have an antecedent and a consequent, as specified below:

BACKGROUND: In its operation, an electromechanical system includes the law of the free fall of bodies as a constant.

CONSEQUENT: **the experiment of the free fall of the bodies has changed its purpose**.

1.6.2 Logico-formal structure of the hypothesis

The structure of the hypothesis expressed by means of a proposition, when translated into logical language, is expressed asf:

In its operation, an electromechanical system includes as a constant the law of the free fall of bodies: P.

The purpose of the experiment on the free falling of bodies has changed: **Q.**

So, the hypothesis has the following logico-formal structure:

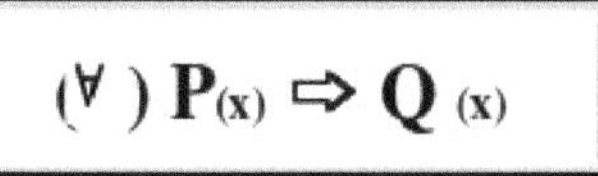

The formula obeys the structure of a logical proposition of molecular type; such algebraic expression, in grammatical language, is paraphrased como as follows: "for any object ***x***, if ***x*** has property **P**, then ***x*** has property Q" (Piscoya, 2007, p. 266).

CHAPTER II

TECHNOLOGY AS AN APPLICATION OF SCIENCE

2.1 The complementarity between technology, science and technology

Logic and mathematics, as formal sciences, cannot be used directly by members of human societies, or, in other words, they do not provide practical benefits directly to social groups. Often, the benefit of the formal sciences is obtained indirectly, the mediating entity being, in this case, technology. Hence, for many epistemologists, technology is described as the result of the application of science. This qualification criterion is not recent, thus, if we appeal to a brief historical reference, we find that the Italian inventor Leonardo da Vinci, when referring to the complementarity between mechanics and mathematics, metaphorically stated that "mechanics is the paradise of the mathematical sciences, because it is thanks to mechanics that we reap the fruits"[3] (Thuiller, 1988, cited by Guevara, 2017, p. 39).

Both science and technology - as fruits of human creativity - try to answer specific questions. Thus, when faced with problems arising from the various gaps in knowledge, science often tries to answer questions such as: "Why", while technology seeks to answer questions such as: "How". For example, a scientific question would be: "Why does a material have a certain limit of resistance when subjected to an external force"; while technology seeks to explain the how, for example: "How to make a battery of a certain make and model perform more efficiently?

Science, when faced with unsolved scientific problems, sometimes finds its answers by itself, other times it makes use of experimental processes, so that, in these cases, it relies on technology and technology; technology, on the other hand, in certain cases also finds its answers by itself, other times it must rely on science -for example, in mathematical calculations-, or even sometimes it relies on technology. A case in which technology relies on science is, for example, when the technologist seeks to make a certain type of machine more efficient, he will - in order to achieve his goal - rely on calculation, for which he works with mathematical data in order to obtain mathematical data by means of which he can communicate his results.

In the case of digital technologies, the technologist will rely on digital logic, for example; this is why we agree with the point of view of the author of the following definition:

> Technological innovation is a direct result of the application of scientific knowledge. Therefore, a policy aimed at promoting innovation and technological development in a country must begin by promoting scientific development, because once we have a high level of scientific knowledge we can apply it to the development of technological innovations, and these in turn will allow industries to become more competitive and, consequently, improve the country's economy, social welfare, etc. (Quintanilla, 2005, pp. 89-90).

The source just quoted gives us to understand that science is not only characterised by a desire to know nature, but also by the will to master it, for which, in many cases, the use of technological means is necessary; in a similar number of cases, the application of science allows the emergence of technological means aimed at solving not only purely scientific problems, but also social ones; in other words, "technology cannot come about without science, since it is nothing but science applied to

[3] Original language of the source: la mecanique, est le paradise des sciences mathematiques, car c'est grace a elle qu'on en recueille les fruits

practical ends" (Bunge, 2012, p. 52). 52); to cite a few examples from contemporary history: electricity, the steam engine, science-based medicine and many other discoveries and innovations have contributed to a considerable improvement in the living conditions of human beings, and this has contributed greatly to the prestige of science (Echeverria, 1999, pp. 250-251).

It should be pointed out that, at present, "technical invention has been almost entirely replaced by technological research, which uses the results and methods of science for utilitarian purposes" (Bunge, 2012, p. 73); along the same lines of thought, as regards the practical utility of technological products, Piscoya considers that

> One aspect that cannot be overlooked when addressing the issue of technology is the social connections it has. In principle, the fact that technological research increases our information mainly in one sense directly satisfies the need for mastery of the natural and the social. As already indicated above, the result of this kind of research teaches us how to do something efficiently and under optimal conditions. This is the reason why the influence of technology on social life, on immediate experience, is enormous, to such an extent that we can affirm that the contemporary world is essentially a product of theoretical science through technological materialisations (1999, p. 173).

In this respect, we agree with the philosopher Piscoya, and we also agree that, just as "philosophy does not directly satisfy any material need, although it could indirectly contribute to it" (Piscoya, 1999, p. 173), in the same way - as we have mentioned above - science, in the strict sense of the word, cannot contribute directly to the welfare of societies. 173), in the same way -as we have mentioned before- science, in the strict sense of the word, cannot contribute directly to the welfare of societies; but, if we accept that technology is an application of science, then we conclude that formal sciences contribute to social welfare, having technology as an intermediary material entity; thus, technology is the result of the application of science, and, of course, social welfare is the result of the application of technology in social reality (in most cases, in today's societies).

It should be borne in mind that, at present, the greater or lesser production and use of technology - even - is a macroeconomic indicator, such that:

> The most immediate differences between developed and underdeveloped countries are in terms of the greater or lesser use of technologies, which are oriented towards greater production and the greater well-being of the societies that create them (Piscoya, 1999, p. 173).

2.2 The language of technology

In general, human knowledge is transmitted by means of language, i.e. language is the medium through which information is transmitted between individuals or between societies. It should be noted that human language is complex and often depends on the code and channel used to establish a certain type of communication.

Human language is complex, because the reality in which each human individual fulfils his or her life cycle is complex and, being included in this reality, human language, the complexity is there. As a way of coping with this complexity, human beings have segmented general human language, assigning certain peculiar characteristics to each of these language segments. Thus we have the general language understood in a broad sense, and the specialised or artificial languages that correspond to only certain segments of reality; humans have elaborated this segmentation out of necessity, since not all of us can encode and decode all types of artificial languages.

Thus, for example, a chemical engineer will not be able to explain a chemical reaction using the musical notes; a musical composer will not be able to write a score using the symbols of the chemical elements contained in the periodic table. Each sector of reality has its own particular characteristics, therefore, each artificial language also has its own characteristics.

As an entity immersed within this complex reality, we find technology; therefore, it is nothing more than another segment of reality; from this we infer that to this type of reality corresponds a language with its own peculiar characteristics: the language of technology.

From an epistemological point of view, the language of technology is characterised as prescriptive, rather than descriptive; that is, technology is fundamentally concerned with the efficient solution of practical problems arising from social needs. The satisfaction of these needs constitutes the goal that gives meaning and direction to the technological prescription that recommends a course of action by virtue of the benefit it produces (Piscoya, 1999, p. 170).

2.2.1 The formal language of technology

To explain this point, we begin by resorting to an example whose statement follows the formal rules of logic. The logical structure of the following scientific law: 'if a gas is heated, then it becomes liquid' is (at the propositional level) if ***A then B***, which can be symbolised as A B. On the other hand, the form of the corresponding nomopragmatic statement is **B *per* A**, which we read 'B by means of Ao 'to obtain B use means A'. The consequent of the law has become the antecedent of the rule; rather, the logical antecedent is now the means, and the logical consequent is now the end of the rule related to the means. The value of the statement A B depends only on the truth values of the atomic propositions that compose it: **A** and **B**: it is a truth function or extensional construction. On the other hand, **B *per* A** is neither true nor false, but effective or ineffective (Bunge, 2012, p. 64).

The contrast between laws and rules can be revealed by recalling the table of values of the conditional **A B** and introducing what we will call the table of effectiveness of the rule **B *per* A.** We will use the sign '1' to designate truth and the sign '0' to designate falsehood; the same signs will serve to represent effectiveness and ineffectiveness, respectively; and the question mark will designate our uncertainty about the effectiveness of the rule (Bunge, 2012, p. 65).

As we can see, in the case of the arithmetic table, we operate with 1 and 0 (binary code); on the other hand, in the case of the efficiency table, a grammatical sign is added to the binary code (the question mark: "?") as shown in the following table:

Table N° 1: *Arithmetic table vs. efficiency table.*

ARITHMETIC TABLE OF THE LAW **A B**			TABLE OF EFFECTIVENESS OF THE **B *per A*** RULE		
A	B	**A B**	A	B	**B *for A***
1	1	1	1	1	1
1	0	0	1	0	0
0	1	1	0	1	?
0	0	1	0	0	?
Formal language of logic			Formal language of technology		

Source*: Bunge, 2012, p. 65*

With regard to the table we have just presented:

Note that the conditional **A B** is only false when the antecedent is true and the consequent is false. On the other hand, the only case in which **B** ***per A*** is certainly effective is when both the means **A** and the end **B** are present. If we do not put the prescribed means into practice and the end is not achieved, or if the end is achieved without using the rule, we cannot know whether the rule is effective or not; only if the end is absent do we know that the rule is ineffective (Bunge, 2012, p. 65).

After analysing the explanation provided by the previous quote, we observe that the language of technology is constituted by rules of action which, according to the proposal of the Argentine philosopher Mario Bunge, have the structure **"A for B"**, that is to say, **"B by means of A"** or **"to obtain B, A must be done"**. This scheme reflects the instrumental sense of the technological rule (**A** is an instrument, a means, to reach **B**, which is the end) (Piscoya, 1999, p. 170).

Obeying the structure of the scheme proposed by Mario Bunge, the technological system that we have invented, in which the law of free fall is applied, is explained as follows

A: is an instrument: invention, i.e. technological innovation.

B: is the purpose: electricity generation.

However, we note that there is a certain incompleteness in this nomopragmatic scheme, since this scheme does not include the social circumstances in which a need can be generated that calls for a solution through the invention of certain technological devices, so we share the point of view of the author of the following quote, who in reference to the scheme proposed by Bunge refers:

But it is insufficient to account for its prescriptive normative meaning. We should not omit the initial circumstances in order to understand the prescribed content in particular situations. We have proposed as a general scheme of technological rules: "In X circumstances Y must be done for ***Z***" (Piscoya, 1999, p. 170).

The general scheme of the technological rule, which we have just quoted, is undoubtedly more complete than the scheme proposed by Bunge, since in its structure it includes an additional indicator, and no less important: the circumstances or context that condition the development of a certain type of technology that allows obtaining the final product or benefit coming from the elaboration of said technology. Therefore, in the scheme proposed by the epistemologist Luis Piscoya, our invention, object of description in the present thesis research, will be explained asf:

In circumstances X: Absence of electricity; use of non-renewable energy sources, etc.

To be made Y: An invention that transforms the gravitational potential energy of water into electrical energy.

To achieve Z: Electricity supply, use of renewable energy sources, reduction of the greenhouse effect.

As can be seen, if the criterion referring to **the initial circumstances** is included in the scheme, the explanation of the technological rule is much more complete and clear, and therefore presents a clearer and more complete epistemic support.

So, in synthesis, we can affirm that the technological rule teaches us to do something efficiently and in optimal conditions; this is the reason why the influence of technology in social life, in immediate experience, is enormous, to such an extent that we can affirm that the contemporary world is essentially a product of theoretical science through technological materialisations (Piscoya, 1999, p. 173).

2.3 The law of free fall of bodies as applied to hydroelectric power plant technology

Undoubtedly, one of the pathetic cases in which we can observe the application of the law of the free fall of bodies is in the generation of electrical energy from the use of the gravitational potential energy of water. In this case, the technology used behaves as the entity which allows the transformation of the force of the earth's gravity into a service of indispensable utility for society: electrical energy.

In Peru there are several installations of this type of technology, a fact that confirms that the Peruvian orography offers certain suitable characteristics in which scientific and technological knowledge can be applied to obtain products that play a preponderant role in the progress of today's societies.

In order to exemplify a technological case in which the law of the free fall of bodies is applied, the following is a brief description of the technological and scientific aspects of a hydroelectric power plant.

2.3.1 Basic Technological Aspects of Hydropower Plants

As we have already mentioned above, at present, there are many types of technologies in which the law of the free fall of bodies is applied; likewise, we have also previously stated that technology is an application of science. At this point, and as an illustrative example, we present the basic components of the technology related to hydroelectric power plants, specifically those using Pelton turbines.

It is important to point out that the operation of this type of technology requires mathematical calculations that include the force of the earth's gravity as a constant and that is part of the formula that explains the operation of a hydroelectric power plant.

Figure N° 1: *Basic components of a hydropower plant.*

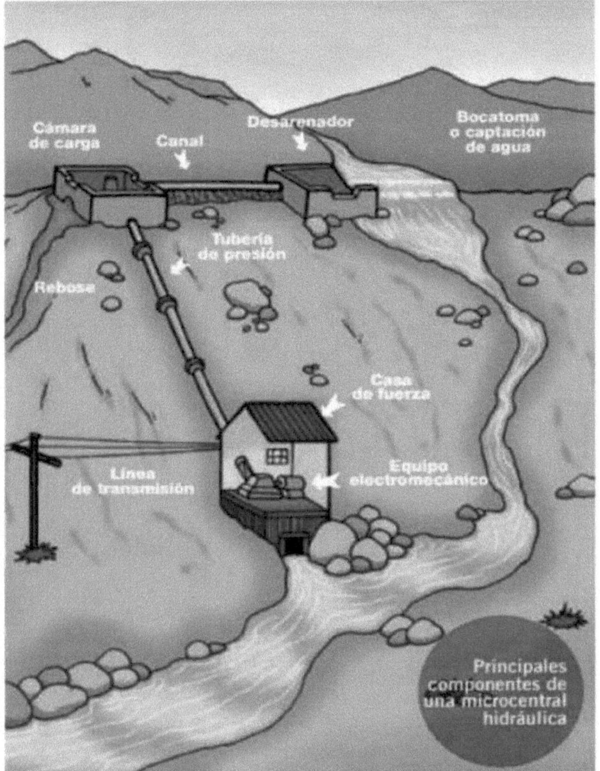

Source: *Practical Solutions.*

From this figure we can infer that the basic components of a hydroelectric power plant, as far as technology and technology are concerned, are the following:

Intake or water intake

Desander

Channel

Charging chamber

Overflow

Pressure pipe

Powerhouse

Electromechanical equipment

Transmission lines

In summary, a hydroelectric power plant consists of: civil works, electromechanical equipment and transmission and distribution networks.

Photo N° 7: *Installed hydroelectric power plant.*

Source: *Practical Solutions.*

2.3.2 Scientific Aspects

Regarding the complementarity between practice and science, it is necessary to take the words of Leonardo Da Vinci who, in his time stated that "those who devote themselves to practice, without taking science into account, are coтo sailors who get into a ship without a rudder and without a compass, and who, therefore, never know where they are going"[4] (Thuiller, quoted by Guevara, 2017); in this regard, a reasonable argument about the proper complementarity between practice and science ıs found ın

The method that Galileo created, the experimental method, consists of a process which, as can be seen, combines a mathematical, rationalist character with an empirical character. By means of it, the aim is to make numbers correspond to phenomena; it is, in a word, to make all phenomena measurable. Only that which can be measured possesses the characteristics of a fully scientific

[4] Original source language: ceux que s'adonnent a la pratique sans la science sont comme des marins qui s'embarquent sans gouvernail et sans boussole, et qui ne saventjamais ou vont.

endeavour, since it combines the mathematical with the empirical, deduction with induction (Galileo, 1984, p.17).

Since the mass of water is a body that is subject to the earth's gravitational force, and since water is also a body that, when it is allowed to fall, the phenomenon of falling bodies is verified, then its respective conceptualisation in mathematical language corresponds to it; and if this mathematical conceptualisation is obtained after the water - on descending - passes through the different technological components of a hydroelectric power station, this phenomenon will have its respective mathematical expression. The mathematical conceptualisation that describes the artificial phenomenon that occurs from the moment the water enters the different accessories that make up the hydraulic technology until the liquid is discharged, has an algebraic form. Once the flow activates the electromechanical accessories, it produces a force called power (P); this power has its respective algebraic expression, which, in its structure, contains its respective variables and constants, as detailed below:

The power that can be obtained from the water resource is estimated using the following equation:

$P = pgQH$

Where:

"p" is the density of water and for calculation purposes is taken as 1000 kg/m^3 , *"g"* represents the acceleration of gravity with a value of 9.8 m/s^2 , *"Q"* represents the flow of water in m^3 /s that is diverted to the turbines and *"H"* is the height or distance measured between the top and the bottom of a waterfall, this distance is also known as gross head and is expressed in metres (Blanco, 2012, p. 36).

The algebraic expression explaining the functioning of the technological system of a hydroelectric power plant is the scientific basis of this system. It is evident that the algebraic expression includes in its structure - among the variables and constants - the acceleration of gravity (*g*).

2.4 Physics as a science that explains the phenomenon of the free fall of bodies.

Physics is a science that is in charge of explaining some of the phenomena that occur in nature; it does so by relying on some sciences that are part of general physics, such as: mechanics, hydrostatics, hydrodynamics, quantum mechanics, etc.

In general, physics as a science is the natural science that studies matter, those general properties that - in the first instance - can be perceived by the senses of the entity it studies: the human being. A researcher of nature, from the perspective of physics, does so on the basis of sensory data provided directly and naturally by his senses (the sense of sight, hearing, smell and touch). Although when the senses reach a limit, the researcher makes use of artificial means provided by technology, such as the telescope, the microscope, etc.; only in this way, relying on exogenous means to his corporeality, does he come to know phenomena and properties of nature; because the technological means allow him to prolong his capacity for sensory perception.

In our research, we use the science of physics to explain mainly mechanical phenomena, particularly those related to action at a distance, i.e. the effect of the earth's gravitational force on bodies on or near the surface of our planet; we therefore rely primarily on the principles of mechanics.

Indeed, as a branch of general physics, classical mechanics explains the motion of matter, in this case by means of a branch of mechanics: dynamics; at the same time,

mechanics also explains matter in the absence of motion: statics.
Now, the artificial Hsic phenomena that are fulfilled in the functioning of the invented mechanism will be explained on the basis of the principles of the so-called classical mechanics, especially the principles related to the phenomenon of the free fall of bodies, in the Galilean perspective.

CHAPTER III

SOME GALILEAN THEORETICAL BASES

3.1 Free fall of bodies

Galileo's scientific concern was motivated by the need to find the truth of a physical phenomenon, more precisely, the Italian scientist pursued the truth about the phenomenon of the fall of bodies. In fact, explaining a physical phenomenon such as the free fall of bodies was one of the most important scientific challenges that Galileo took on. The difficulty was enormous. Let's bear in mind that today, to adequately study this type of movement, it is necessary to resort to technology, such as instant photography, which did not exist in his time. Objects fall too fast and precision instruments are needed to study them properly. Galileo overcame this difficulty in a most elegant way: he studied this motion by using the inclined plane, which was a way of 'circumventing gravity' and achieving an equivalent experiment that could be studied. A plane whose inclination becomes greater and greater, in the limit will have a vertical direction (Cork, 2012, p; 111).

Galileo made use of a method of time measurement based on a water clock. He marked the position of the dial on the inclined plane at equal intervals of time. From these marks, Galileo realised that the distances travelled during the time intervals kept an odd-numbered proportion: 1, 3, 5, 7. Since the proportions held with more inclined planes, this same effect had to occur in free fall. The time taken to traverse each unit of space is 1, 3, 5, 7..., which means that it takes one unit of time to traverse the first section; at the end of the second section it has taken a total of 1+3 = 4 units of time (Corcho, 2012, p. 111).

This source describes one of the scientific contributions explaining the free fall of bodies due to the earth's attractive force; on the subject in question:

One of the most familiar cases of constant acceleration is due to gravity near the Earth's surface. When an object falls, its initial velocity is zero (at the instant it is released), but some time later during the fall, it has a non-zero velocity. There has been a change in velocity and, by definition, an acceleration. The acceleration due to gravity (g) has a value of approximately 9.80 m/s^2 . Objects moving only under the influence of gravity are said to be in free fall. The acceleration due to gravity g is the constant acceleration for all objects in free fall, regardless of their mass and weight (Astorga, 2010, p.43).

Some of the extracts that give theoretical and scientific support to our thesis research are taken from Galileo's *Dialogues on Two New Sciences*, such as the one quoted below:

Therefore, when I observe that a stone descending from above from rest, acquires new increments of velocity, why should I not believe that such increments occur in the simplest and most obvious proportion? Now, if we look at it, we shall not find any increase or increase simpler than that which always increases in the same way. What we shall easily understand by considering the close relation between time and motion[5] (Galilei, 1980, p.79)

The same author, in another section of his work mentions: We call uniformly accelerated motion that which, starting from rest, acquires equal increments of speed during equal times (Galilei, 1945, p. 216).

[5] Original language of the source: Quando, dunque, osservo che una pietra, che discende dall'alto a partire dalla quiete, acquista via via nuovi incrementi di velocita, perche non dovrei credere che tali aumenti avvengano secondo la piu semplice e piu ovvia proporzione? Now, if we consider things carefully, we will not find any increase or increase more simple than that which always increases in the same way. Which we can easily understand considering the close connection between time and motorbike.

THEOREM II - PROPOSITION II

If a mobile with uniformly accelerated motion descends from rest, the spaces traversed by it at any times are the squared ratio of the same times, i.e., the squares of those times (Galilei, 1945, p. 225); in short, we have that:

Galileo established experimentally the laws of the fall of bodies which are at all points in opposition to the peripatetic belief. He thus introduced the absolutely new concept of 'acceleration'. He showed that the velocity of falling bodies is the same for all, whatever their weight; that it is not proportional to the space travelled but to the time employed ($v = gt$); and he established the law according to which the spaces travelled by the falling body are proportional to the squares of the times during which they have been travelled ($s = 7\, gt^2$). He showed that the air does not increase the speed of the fall but, on the contrary, decreases it, and that the movement of the fall in the void would not therefore be naturally uniform but naturally accelerated (Garcia, 1957, p. 144).

3.2 The scientific-experimental method of Galileo Galilei

According to many historians, epistemologists and scientists, Galileo was the one who initiated experimentation in science, since he had the ability to manipulate the variables in his process of experimentation; and the results of this research he presented them expressed in mathematical language; that is, mathematization understood as an operational definition. In the introduction to the Spanish language edition of Galileo's work *The Assayer*, it is mentioned that

The method that Galileo created, the experimental method, consists of a process which, as can be seen, combines a mathematical, rationalist character with an empirical character. By means of it, the aim is to make numbers correspond to phenomena; it is, in a word, to make all phenomena measurable. Only that which can be measured possesses the characteristics of a fully scientific endeavour, since it combines the mathematical with the empirical, deduction with induction (Galileo, 1984, p. 16).

This means that, if we seek the truth of natural phenomena, the questions that are asked of nature, and the answers that are obtained from those questions, must be expressed in mathematical language, as Galileo Galilei himself stated in 1623:

Philosophy is written in that great book which is open before our eyes, that is to say, the universe, but it cannot be understood unless one first learns to understand the language, to know the characters in which it is written. It is written in mathematical language and its characters are triangles, circles, and other geometrical figures, without which it is impossible to understand a word; without them it is like turning vainly in a dark labyrinth (1984, p. 61).

In this way, the Italian scientist emphasised the importance of mathematics when it comes to knowing the truth of natural phenomena and, at the same time, he made it clear that mathematics is an inherent property of nature; that is, mathematical properties are an intrinsic quality of natural phenomena; therefore, if we wish to "communicate" with nature, in order to receive accurate and truthful information, it is necessary to learn this unspoken language of the universe: mathematics, which, in itself, is a type of language that nature has only "written", but which is sufficient for us to know the truth of the phenomena. Together with this Galilean contribution, of a rational nature, there is the experimental facet, that is to say:

There are his concrete discoveries and there is what could be called 'the miracle of Galileo': his having created with these still inadequate and imperfect materials a new and comprehensive system, with a new method of reasoning and experimentation, a system and method that are prolonged and subsist in our modern science (Garcia, 1957, p. 143).

Therefore, we can affirm, like so many other researchers, that modern natural science was born with the scientist Galileo Galilei, who:

He is not satisfied with pure (theoretically neutral) observation or arbitrary conjecture. Galileo proposed hypotheses and put them to experimental test. He thus founded modern dynamics, the first phase of modern science. Galileo is keenly interested in methodological, gnoseological and ontological problems: he is a scientist and a philosopher and, in addition, an engineer and a language artist (Bunge, 1976, p. 30).

CHAPTER IV

TRANSFORMATION OF THE EARTH'S GRAVITATIONAL FORCE INTO ELECTRICAL ENERGY BY MEANS OF A CONTAINER SYSTEM

4.1 Practical application of the law of the free fall of bodies in an electromechanical system

According to the scientific principle of the law of energy conservation or the first law of thermodynamics cited by Earls: "Energy can neither be created nor destroyed" (2007, p.87), then, as soon as the invented mechanical system comes into operation, the energy coming from the earth's attraction force is transformed into electrical energy, mainly thanks to a mechanical mediating entity. This mediating entity is the system of containers that has been invented in order to obtain electrical energy, taking as a source the gravitational potential energy of the water contained in containers which are accessories that are part of the electromechanical system in general.

In order to explain the methods and procedures that we have followed to carry out the practical application of the law of the free fall of bodies in an electromechanical system, we will now present the basic details of the experimental process:

4.2 Aims of the experiment

- Transforming the gravitational potential energy of water into electrical energy.
- To invent a technological instrument to transform the gravitational potential energy of water into electrical energy.

4.3 Hypothesis of the experiment

- It is possible to transform the gravitational potential energy of water into electrical energy.
- Since the invention of a mechanical instrument it is possible to transform the gravitational potential energy of water into electrical energy.

4.4 Variables, constants and factors involved in the experiment

4.4.1 Independent variables

Flow rate

Height

4.4.2 Dependent variables

Mechanical power

Electrical power

Weight

Linear speed

Angular velocity

4.5 Constants

Earth's gravitational force

Air density

Atmospheric pressure

4.6 Intervening factors

Air humidity

Wind speed and direction Ambient temperature

4.7 Brief description of the mechanical system used in the experiment

In order to carry out the various experimental steps and processes, we developed a mechanical instrument which, to a certain extent, fulfils the functions of a hydraulic turbine, in that it generates a propulsion which is then transmitted to the generator; this propulsion is generated thanks to the potential energy of the water. As we will see below (Figure N° 2), the mechanical system mentioned above is not similar in shape to any type of conventional turbine, i.e. it does not have a circular shape, but has the shape of a closed belt with two pulleys at the ends, i.e. it is a transmission belt which generates propulsion; it is because of this propulsion that the vessel system resembles a hydraulic turbine to a certain extent.

Figure N° 2: *Basic scheme of the invention experimented with.*

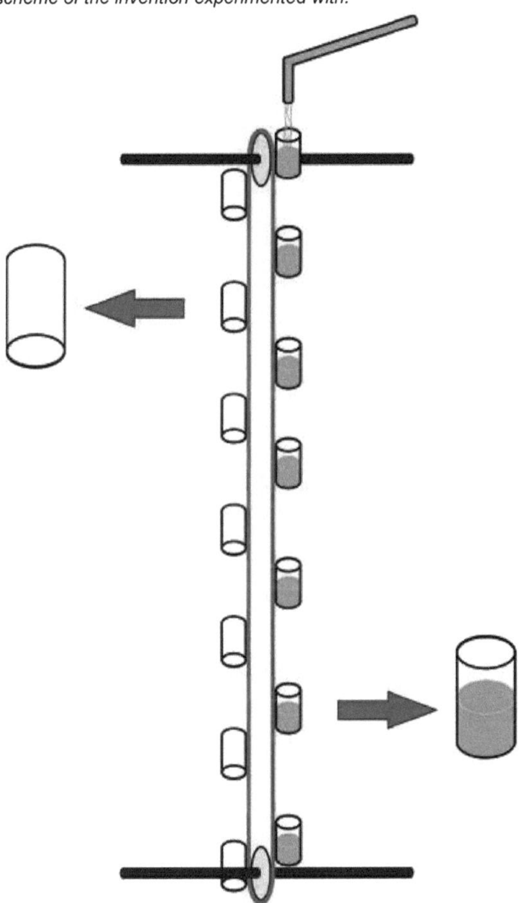

Source: *Own elaboration.*

The shape of the invention shown in figure N° 2 was designed for didactic purposes, since in the experiment we will use 48 containers (at this stage: glasses) coupled to a closed belt; as we can see in the figure, the glasses located on the right contain water, then, because of their weight and due to the effect of gravity, they generate the propulsion in the belt, making the two pulleys rotate immediately and, with this, the other accessories of the electromechanical system, including the generator, start to work.

So, in the experiment, we have 24 empty vessels on one side of the belt and 24 full vessels on the other, thus generating constant linear and circular movements observed in the belt and the two pulleys on which the belt runs.

As for the link between the scientific and technological parameters, the belt is the mechanical accessory that transmits the force of gravity to the generator via shafts, pulleys and a smaller transmission belt as shown in the following picture:

Photo N°8: *Electromechanical accessories used in the experiment.*

Source: *Own elaboration.*

VALUES OF THE VARIABLES BELONGING TO THE MECHANICAL SYSTEM IN THE EXPERIMENT.

Height (in metres)	**8,56**
Volume (L/min)	**10,757**

VALUES OF VARIABLES THAT BELONG TO THE GENERATOR IN THE EXPERIMENT.

Variables	According to the manufacturer	**In the experiment**
Volt DC	12	**11,56**
rpm	420	**450**

Let's look at some figures on the measurement of dependent variables, Kpics of the generator:

Photo N° 9: *Experimental data collection instruments.*

Source: *Own elaboration.*

The data provided by the instruments are:

AC voltage	10,27
DC voltage	12,00
rpm	490

Photo N° 10: *House in which we installed the various electromechanical accessories needed to carry out the experiments.*

Source: *Own elaboration.*

4.8 Conclusions of the experiment

FIRST: By using a belt as a means of transmitting gravitational potential energy, with a flow rate of 10, 757 L/min, at a height of 8.56m, we have managed to obtain 11, 56 volts DC.

SECOND: By making use of an invented technological instrument it is possible to transform the gravitational potential energy of water into electrical energy.

4.9 Patent of utility model: "Vessel system for electricity generation".

As we have mentioned in different previous sections, the present thesis research describes and explains the methods and processes that made possible the application of the law of the free fall of bodies in an electromechanical system, which, with a certain flow rate and at a suitable height, allowed the generation of electrical energy.

4.9.1 Background to the patent

In order to obtain a scientific law to explain the free fall of bodies, Galileo -among other mechanical devices- used the inclined plane; in our case we basically used a system of containers in order to obtain electrical energy; However, like all research, it had to go through various processes, including experimental processes, which is why at the beginning we used a mechanical system that, although it did not meet the technical and legal characteristics to be patented (see figure N° 2); however, this prototype was used to carry out different experimental stages *in situ*, some of whose results have already been exposed in the previous chapters of this research.

TECHNICAL STRENGTH OF THE PROTOTYPE: It allowed to prove that it is possible to generate electrical energy with heights and flows with which other technologies do not work, specifically Pelton turbines.

TECHNICAL WEAKNESS: A belt was used as a mechanical means to generate the propulsion and on this a single column of containers was coupled; on the other hand, at the moment of the activation of the whole electromechanical system, the water came into contact with the belt and the pulleys on which the belt circulated, so its durability was short-lived and it lost tension.

TECHNICAL-ADMINISTRATIVE WEAKNESS: When searching for similar prototypes with a view to patentability, the specialists concluded that it did not meet the novelty requirement (see annex N° 8).

4.9.2 Patent: 'Electric Power Generation Vessel System'.

In order to improve the prototype that was used to develop the different experimental processes, different technical improvements were made, which turned out to be significant.

TECHNICAL STRENGTHS OF THE PATENTED PROTOTYPE: The innovation was such that the new prototype had more than one column of containers, i.e. the new prototype had at least two columns of containers, as shown below:

Figure N° 3: *Basic scheme of the vessel system.*

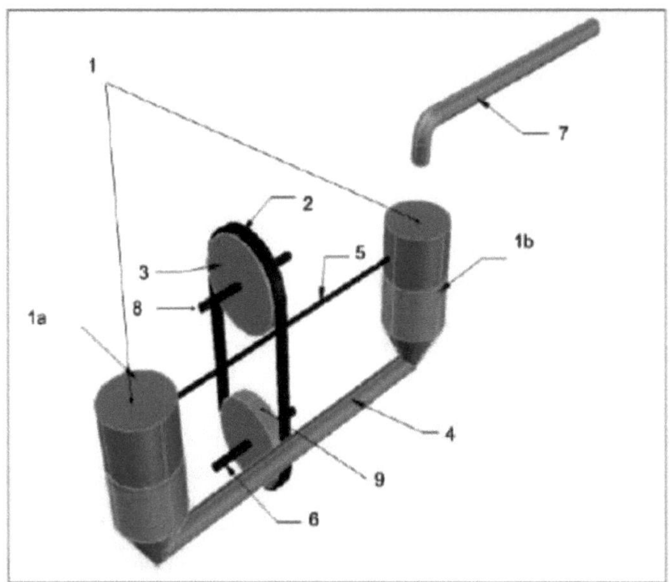

Source: *Indecopi.*

The fact that there are two columns of containers makes it possible to accumulate more water per unit of time, significantly increasing the mechanical power and thus the electrical power as well.

It should be pointed out that the mechanical system referred to in our invention allows greater power to be obtained in relation to conventional turbines; in other words, with the same flow rate and the same head, the electromechanical system using the chain and containers allows greater mechanical power to be obtained and, therefore, greater electrical power, as it operates with a much reduced flow rate of around 10^{-4} m^3 /s, in relation to conventional turbines. It should be noted that hydraulic turbines have some technical limits, such as, for example, the volume required for efficient operation; for example, the most widely used turbine, the Pelton turbine, operates at a minimum flow rate of 5 x 10-2 m^3 /s.

For greater efficiency and durability of the mechanical accessories, the use of the belt was discarded to be replaced by a steel chain used in industrial technology as shown in photo N° 11, in the same photo you can see the model that we exhibited in the XVII National Competition of Inventions and Industrial Designs, organised by Indecopi, in November 2018, in which event our invention was exhibited representing the Universidad Nacional Mayor de San Marcos.

Photo N° 11: *Model showing the two chains and three columns of containers.*

Source: *Own elaboration.*

Undoubtedly, the use of chains makes possible a greater resistance to the weight, as well as a greater durability of the mechanical system, at the same time, it makes possible to obtain a higher power. Also, as we can see in the image of the model, the modification presents other advantages in addition to the previous ones: it is possible to use two, three, four, or more columns of containers, which allows to increase considerably the mechanical power.

Figure N° 4: *Graphical comparison.*

Source: *Indecopi.*

Another advantage we can highlight is that the patented prototype allows the containers to be located at a strategic distance from the chain, and the water enters only through one of them, from which it is distributed to the other containers by the principle of communicating vessels. The distance between the containers and the chains in this case means that the water does not come into contact with either the pulleys or the chains, which favours the lubrication process and thus guarantees the durability of the mechanical system.

SCIENTIFIC-EXPERIMENTAL WEAKNESS: Although we have been successful in the patenting process, it is necessary to clarify that we have not yet carried out experimental processes *in situ* with the patented prototype. Following the rules on intellectual property of the UNMSM, and with scientific objectives, we hope that in the shortest possible time we will be able to carry out the different experimental stages and then its application with the aim of achieving the social welfare of those populations located in places where the basic technical conditions exist for the application of this technology.

4.9.3 Obtaining a patent

Once we presented the technical document about the invented mechanical system before the corresponding instance of the Indecopi, after an exhaustive evaluation, the evaluators concluded that the invention does comply with the technical-administrative requirements, such as passing the patentability examination and passing the search report (see ANNEXES: 5 and 6). One of the results of the evaluation is the following graphic comparison:

Figure N° 5: *Graphical comparison.*

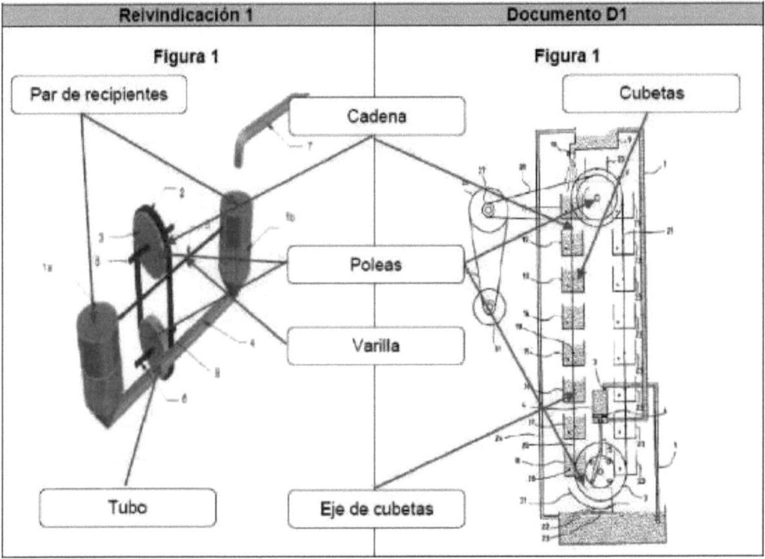

Source: *Indecopi.*

After a little more than two years of administrative procedures, the Indecopi granted us the patent on 05-11-2020 (see annex N° 7).

The technical and scientific information of this invention is detailed in the document that was drafted following the rules established by the Indecopi, which has the legal denomination of *Technical Document*, this document was drafted and presented to the Indecopi, as part of the process of registration of intellectual property of the technological innovation in reference.

In order to present the technical and scientific details of the invention, we have deemed it necessary to attach to the present thesis research the same *Technical Document* that was filed before the Indecopi (see ANNEX N°1); therefore, in this section we limit ourselves to present a summary, which is also contained in the aforementioned document:

It is known that the technology that allows the production of hydropower is generally made up of various types of turbines; however, these operate at certain heights and with certain flow rates, i.e., they have a Kmite. To overcome these disadvantages, other ways of harnessing the potential energy of water have been devised; among these are machines that allow the potential energy of water to be harnessed through the use of two pulleys joined together, either by cables, belts or endless chains, in all three cases, with buckets or containers attached to the belts, cables or chains. The disadvantage is that the durability period of the mechanical accessories mentioned is short, mainly because the water is in permanent contact with these mechanical elements, especially with the pulleys and the belts, cables and chains, as the case may be. In view of this situation, a technological innovation was made, which refers to the use of pulleys with a chain, on which at least one pair of containers is attached on each side of the chain, i.e. one pair loaded with water and the other empty. Each container is located at a distance from the side of the chain, so that when the water is loaded and unloaded, it does not come into contact with either the pulleys or the chain, thus allowing the chain and pulleys to be lubricated, which means that the accessories in question have a longer service life. In the case of high flow rates, it is possible to attach a second chain, and with this, three containers can be attached, this fact allows to accumulate more water for each unit of time and distance, which implies the increase of the mechanical power (UNMSM, 2018, p. 13).

CONCLUSIONS OF THE THESIS RESEARCH

FIRST: After having carried out a theoretical and experimental analysis linked to the phenomenon of terrestrial attraction, we can affirm that the experiment of the free fall of bodies is susceptible to change its purpose; our study confirms that the teleology of this experiment bifurcates, each time that some scientist, technologist or inventor finds an additional application to those already known.

SECOND: In view of the fact that we have been the promoters of one more technological application of the law of the fall of bodies, we hypothesize that there are many other technological and scientific applications that could solve certain problems that afflict society; but these hypothetical applications have not yet been discovered, therefore, we do not know with certainty what or how many they are, nor when they will be discovered and put into practice.

THIRD: The application of the law of the free fall of bodies, application of which we have been managers, in the course of time, is neither the only one nor the last one, but it is a pathetic example of the change of teleology of the experiment of the free fall of bodies, that is, it is one more bifurcation of the teleology made reality in a time and space in which this thesis research is framed; likewise, this bifurcation constitutes an alternative solution to a social problem.

FOURTH: The explanation of the process of generation of electric energy, taking the mass of falling water as its source, scientifically depends on the law of free fall discovered by Galileo Galilei.

FIFTH: Galileo Galilei used the inclined plane to solve a scientific problem; on the other hand, in our case, we use a chain, pulleys and containers to apply the law of the free fall of bodies to obtain a social benefit.

SIXTH: The interaction *Homo sapiens* sapiens-terrestrial gravity *force*, has always taken place at a technical, scientific and technological level; either to face the terrestrial gravity force or to take advantage of it.

SEVENTH: The Peruvian territory presents some orographic conditions in which certain types of hydraulic technologies can be installed in order to solve certain social problems.

EIGHTH: Galileo Galilei did his duty in discovering the law of the free fall of bodies; our duty is to discover its applications, as many as possible.

BIBLIOGRAPHICAL REFERENCES

Astorga, M. (2010) *Galileo's model of an inclined plane for teaching kinematics* (Master's thesis) Centro de Investigacion en Materiales Avanzados, S.C. Juarez.

Blanco, E. (2012) *Estimation de la potencia electrica teorica disponible en Rio Copinula*, Jujutla, Ahuachapan. Retrieved 29 January 2020, from:http://www.redicces.org.sv/jspui/bitstream/10972/1971/1/2-estimacion- de-la-potencia-electrica-teorica-disponible-en-rio-copinula-iuiutla- ahuachapan.pdf

Bunge, M. (1972) *La investigation cientifica*. Barcelona, Ariel

(1976) *Epistemologia*. Barcelona, Ariel.

(2012) *Filosofta de la tecnologia y otros ensayos*. Lima, Fondo editorial de la UIGV

(2001) ^Que es filosofar cientfficamente? Lima, Fondo editorial de la UIGV

COP25 (2019) *COP25 Chile - Madrid 2019*. Retrieved January 22, 2020, from https://noticiasibex35.com/energia/cop-25/

Corcho, R. (2012) *La naturaleza se escribe con formulas.* Navarra, EDITEC

Deza (2012) *Cajamarca, Cumbemayo, el Camino del Agua*. Lima, UAP.

Earls, J. (2006) *La agricultura andina ante una globalizacion en desplombiacion*. Lima, PUCP

(2007) *Introduction to the theory of complex systems*. Lima, PUCP

Echeverna, J. (1999), *Introduction a la metodologia de la ciencia. La filosofia de la ciencia en el siglo XX*, Madrid, Graficas Rogar S.A.

Fernandez, (2012) *¡Eureka! El placer de la invencion.* Spain, EDITEC

Galileo, G. (1984) *El Ensayador*, Madrid, SARPE.

(1945), *Dialogos Acerca de Dos Nuevas Ciencias,* Buenos Aires, Editorial Losada.

(1980) *Discorsi e dimostrazioni matematiche intorno a due nuove scienze* (Discourses and mathematical demonstrations around two new sciences.) Italy, UTET.

Garrta, E. (1957) *Historia de la fisica.* Buenos Aires, Editorial Nova

Guevara, A. (2017), *La caduta dei gravi: dal movimento locale alla gravitazione universale* [The fall of bodies: from local motion to universal gravitation] (Master's thesis). University of Florence. Italy INDECOPI (2020) *^Who are we?*. Retrieved January 22, 2020, from https://www.indecopi.gob.pe/quienes-somos

Piscoya, L. (1999), *Filosofia.* Lima, Metrocolor

(2007) *Logica General*. Lima. Vicerrectorado academico UNMSM

Quintanilla, M. (2005), *Filosofia de la tecnologia.* Lima, Fondo editorial de la UIGV

Soluciones Practicas (2012) *Micro central Hidroelectricas*. Lima, Soluciones Practicas

UNMSM (2018) *Reglamento de Propiedad Intelectual de la Universidad Nacional Mayor de San Marcos.* Lima, VRIP UNMSM

(2018) *Sistema de Recipientes para Generation de Energia Electrica* (Documento tecnico) Lima, UNMSM

Vollmer, G. (2005), *Teona evolucionista del conocimiento.* Madrid, Editorial Comares

ANNEX N° 1

1

SISTEMA DE RECIPIENTES PARA GENERACION DE ENERGIA ELECTRICA

Campo de la Invención

La presente invención se enmarca en el campo técnico de las máquinas que convierten la energía potencial gravitatoria del agua en energía eléctrica, utilizando un sistema de recipientes dispuestos sobre una cadena, en cuyos recipientes se acumula el agua proveniente de una fuente alimentadora.

Antecedentes de la Invención

Existen diferentes tipos de máquinas que convierten la energía potencial del agua en energía eléctrica como las turbinas hidráulicas que, dependiendo de la altura y del volumen, se utilizan para generar hidroenergía, en un rango que va desde las denominadas pico centrales hasta las grandes centrales hidroeléctricas.

Las turbinas hidráulicas presentan algunos limites técnicos, como por ejemplo, el referido al volumen necesario para que funcionen con eficiencia; por citar el de la turbina más usada, la pelton, trabaja a un caudal minimo de 5×10^{-2} m^3/s. Además, la potencia mecánica depende de la presión hidrostática, a su vez, el impacto del agua es con un solo punto tangencial de la circunferencia de la turbina por lo tanto es momentáneo.

Al respecto, la presente invención permite generar electricidad con un caudal muy reducido porque la propulsión se genera por efecto del peso del agua acumulada en los recipientes, por lo tanto, el peso del agua ejerce una fuerza constante, a través de la cadena, sobre parte de las circunferencias de las poleas, esta fuerza se conserva, desde que el momento en que los recipientes se cargan en el momento que giran sobre la polea ubicada en la parte superior hasta el momento en que se produce la descarga en la polea inferior. Adicionalmente la presente invención genera electricidad con un caudal muy reducido del orden de 10^{-4} m^3/s, lo que significa un caudal muy bajo en relación a la turbina pelton; a su vez, en cuanto se refiere a la altura, el presente invento funciona con eficiencia a bajas alturas; así tenemos que al menos funciona con una distancia aproximada de un milimetro de distancia entre las circunferencias de las dos poleas sobre las cuales gira la cadena, lo que permite el funcionamiento eficiente de la invención a bajas alturas.

Anteriormente, el equipo de investigación conformado por docentes y estudiantes de la Universidad Nacional Mayor de San Marcos, en el año 2012 desarrolló una máquina con dos poleas, recipientes y fajas y/o cables que funcionó con los siguientes parámetros: altura de 8.5 m y caudal de 11 l/m, dicha máquina logró accionar un generador con las siguientes características: 12 voltios, 420 rpm, 330W con una eficiencia de 70%, lo que significa haber superado los límites referidos a la relación entre la altura y el caudal necesarios para que funcionen las turbinas convencionales, como por ejemplo, la turbina pelton. Dicha máquina utiliza como elemento mecánico propulsor una faja o cables de driza, sobre cuyos accesorios se adhieren los recipientes; aparte de su corto periodo de vida útil de la faja, poleas y recipientes; la faja, pierde tensión en cuestión de días; al respecto, la presente invención utiliza una cadena de acero en vez de fajas o cables la cual evita la pérdida de tensión a corto plazo y, a la vez, puede trabajar con volúmenes de agua más elevados.

Al respecto, el documento KR20100034129 describe un sistema micro hidroeléctrico de baja altura que impulsa un generador mediante el movimiento de una pluralidad de cubos ubicados en una rueda dentada que se conecta al eje del generador.

Este sistema presenta un inconveniente, ya que el centro de gravedad del recipiente al estar a una cierta distancia de la cadena y al frente de ésta, la fuerza total del peso del recipiente, no se transmite en su totalidad hacia el punto tangencial de la polea ubicada en la parte superior, por lo tanto, al realizar la descomposición de fuerzas, según el diagrama de cuerpo libre de las fuerzas intervinientes, se observa un vector en dirección horizontal.

Sin embargo la presente invención presenta una ventaja ya que el centro de gravedad del par de recipientes coincide con los puntos tangenciales de las polea superior e inferior; dichos puntos son en donde concurren la circunferencia de las poleas y la cadena; por lo que es posible que los recipientes se carguen de agua en la totalidad de su capacidad volumétrica, y, además, ya que el centro de gravedad del par de recipientes coincide tangencialmente con la circunferencia de la polea, la fuerza generada por el peso, es transmitida en su totalidad hacia el punto tangencial en el que concurren la cadena y la polea ubicada en la parte superior. Por lo tanto al hacer la descomposición de fuerzas según el diagrama de cuerpo libre, no se observa ninguna fuerza en dirección horizontal, por lo tanto, el cien por ciento de la fuerza del peso se transmite a la cadena.

El documento WO2013036001, describe un sistema generador de energía que comprende una rueda superior y una rueda inferior formando una línea perpendicular con respecto al suelo, un carril que circula alrededor de la rueda superior y la rueda inferior, y una pluralidad de cubos fijados a intervalos uniformes en el carril; un rotor que tiene una pluralidad de dientes fijados en una superficie exterior del mismo a intervalos uniformes e instalados en acoplamiento con las barras de las cubetas; y un fluido suministrado a las cubetas para ejercer energía potencial en el sistema, donde la energía del fluido se usa para aumentar la fuerza de rotación del rotor que permite generar energía eléctrica.

Este sistema presenta un inconveniente puesto que el agua está en contacto directo con las poleas y el carril, además, el centro de gravedad de los cubos no coincide con la circunferencia de la rueda o polea, por lo que un porcentaje de la fuerza del peso del recipiente no se transmite hacia la polea ubicada en la parte superior.

Por lo expuesto, resulta necesario proveer un sistema que supere los problemas presentados por los antecedentes antes mencionados.

Descripción de la Invención

Con la finalidad de resolver los problemas antes mencionados, se ha ideado un sistema de recipientes que permite transformar la energía potencial del agua en energía mecánica, para luego ser transformada en energía eléctrica mediante un sistema electromecánico. La transformación de energía ocurre a partir del momento en el que la fuerza del peso del agua genera la propulsión que, mediante una cadena, se transmite a una polea superior, cuyo eje está interconectado con un sistema de ejes y poleas que aumentan la velocidad que, en conjunto, hacen girar al rotor del generador.

La máquina en referencia comprende un sistema de recipientes con al menos dos pares de recipientes interconectados mediante una varilla en la parte superior, y un tubo en la parte inferior, este tubo permite que el agua fluya por los recipientes según el principio de vasos comunicantes, por lo que el agua es distribuida entre los recipientes equitativamente; dicho principio establece que si en varios recipientes comunicantes se vierte un líquido homogéneo y se deja en reposo, en breve, se observa que en todos los vasos el líquido alcanza al mismo nivel, independientemente de la forma o de la capacidad volumétrica de los recipientes.

Cada conjunto de recipientes interconectados presenta al menos un par de recipientes los mismos que se interconectan entre si, mediante un tubo en la parte inferior de éstos y una varilla en la parte superior como medio de soporte. Según las características de los parámetros físicos determinantes, tales como el volumen y la altura; los recipientes pueden ser de diferentes capacidades volumétricas, pero en todos los casos, cada uno de los recipientes está hecho, preferentemente, de fibra de vidrio y tiene una forma aerodinámica que le permite romper la resistencia del aire con más facilidad y, con ello, aumentar la potencia mecánica de la máquina. El tubo se conecta con cada recipiente por la parte inferior, en puntos que coinciden con los extremos del diámetro de cada recipiente, mientras que la varilla atraviesa por la parte superior a los recipientes, cercano a la boca de cada recipiente en puntos que coinciden con los puntos extremos del diámetro de cada recipiente. Las dimensiones y tipo de materiales de estos dos elementos de interconexión dependen del volumen y altura disponibles. En el caso de la varilla, preferentemente, es de acero inoxidable, para evitar la corrosón y, a la vez, resistir el peso de los recipientes a los que sostiene y, con respecto al tubo, preferentemente, es de polietileno y no es afectado significativamente por el peso, ya que se ubica en la parte inferior de los recipientes cargados de agua.

La acumulación de la masa, mediante el sistema de recipientes, permite que el agua ingrese solamente por uno de los recipientes, desde dicho recipiente el agua será transmitida hacia los demás recipientes interconectados, esto permite que la distribución del agua sea equitativa y, por ende, el peso de cada recipiente sea igual.

Este tipo interconexión y distribución de los recipientes, permite que los recipientes sean ubicados en la parte lateral de la cadena a una distancia estratégica, logrando, así, que el agua no haga contacto ni con la cadena ni con las poleas, de tal manera que el sistema mecánico ofrece una mayor durabilidad, evitando, así, un desgaste acelerado y, a la vez, al no contactar el agua con la cadena ni con las poleas, facilita el proceso de lubricación.

El sistema presenta al menos una cadena cerrada del tipo sin fin, preferentemente hecha de acero, que puede ser, en cuanto a su estructura, del tipo: simple, doble, triple, etc. y, en cuanto a su tamaño, puede ser cota o extensa, dependiendo del caudal y la altura. Además, el sistema presenta al menos, dos poleas de acero con engranajes en su circunferencia, sus dimensiones dependen de la altura y el caudal disponible. Sobre estas dos poleas circula la cadena.

En la mencionada cadena se acoplan los recipientes interconectados, ubicados en la parte lateral de ésta, de tal manera que los centros de gravedad de cada recipiente se alinean horizontalmente con la cadena y, por ende, también se alinea verticalmente con el punto de tangencia de la cadena con las poleas, al descender y ascender; este hecho permite que la fuerza total producida por el peso se transmita en su totalidad a las poleas y, desde éstas, hacia el sistema electromecánico de generación de energía eléctrica.

La propulsión se inicia en el momento que una cantidad suficiente de agua, mediante un tubo de alimentación, es depositada en los recipientes adheridos a la cadena, la fuerza de gravedad que ejerce la Tierra sobre cada recipiente se transmite a la cadena, entonces la cadena gira sobre dos poleas dentadas que están ubicadas una al extremo de la otra y separadas perpendicularmente; la propulsión generada por el peso del agua es transmitida, finalmente, a las dos poleas, y es la polea superior la que hace girar al eje del rotor del generador.

Los recipientes interconectados, están fijados a la cadena cerrada, de tal manera que en un lado de la cadena, los recipientes mantienen sus bocas hacia arriba y descienden al ser cargados con agua; mientras que en el lado opuesto de la cadena, los recipientes mantienen una posición con sus bocas hacia abajo, pues están vacíos y ascienden hasta la polea superior para volver a ser cargados otra vez, y así, sucesivamente, siempre que la fuente alimentadora de agua sea constante. Como ya se ha mencionado líneas arriba, el sistema de recipientes interconectados permite transmitir la fuerza total del peso hacia la cadena y con ello a las poleas; por otro lado, el uso del sistema de vasos comunicantes hace posible de que, en el momento de la carga y descarga de los recipientes, el agua no haga contacto ni con las cadenas ni con las poleas; además, ya que el agua se acumula en los recipientes ubicados en las partes laterales de la cadena, el sistema de recipientes permite obtener una mayor capacidad volumétrica por cada unidad de distancia y por cada unidad de tiempo, lo que permite generar una mayor potencia mecánica y, por ende, eléctrica.

Todo este sistema mecánico, de cierto modo, hace las funciones de una turbina a reacción; ya que la propulsión emana del peso de la masa contenida en los recipientes, éstos, a su vez, atraidos por la fuerza de gravedad terrestre, hacen que la cadena ejerza una fuerza tangencial sobre las dos poleas que se encuentran en los extremos; a su vez las poleas transmiten el movimiento circular al generador, mediante el eje de la polea superior, cuyos extremos descansan sobre sus respectivos cojinetes que están acoplados a una base que puede ser de madera y/o metal.

El sistema mecánico referido a la presente invención, permite obtener una mayor potencia en relación a las turbinas convencionales; es decir, con el mismo caudal y la misma altura, el sistema electromecánico que utiliza la cadena y recipientes, permite obtener una mayor potencia mecánica y, con ello, una mayor potencia eléctrica, pues funciona con un caudal muy reducido del orden de 10^{-4} m^3/s, en relación a los antecedentes mencionados.

En cuanto se refiere a los recipientes, éstos están ubicados a una distancia estratégica y en la parte lateral de la cadena, además, están acoplados a la cadena en el punto céntrico de la varilla y del tubo que interconectan a los recipientes, de tal manera que haya una distribución equitativa del agua entre los recipientes para que se dé un equilibrio entre ellos, esto permite que en el momento de la carga y descarga del agua, ésta no haga contacto ni con la cadena ni con los engranajes de las dos poleas; de esta manera, de un lado, se evita que el agua ocasione un desgaste rápido de la cadena y, de otro lado, permite que la cadena y las poleas y sus respectivos engranajes conserven su lubricación según las especificaciones técnicas del fabricante, garantizando, así, el periodo de vida útil de todo el sistema mecánico de la máquina.

El centro de gravedad del par de recipientes coincide con el punto tangencial en donde concurren la cadena y las poleas, hecho que permite que el sistema funcione con más eficiencia; por lo que es posible que los recipientes se carguen de agua en la totalidad de su capacidad volumétrica. La fuerza generada por el peso, es transmitida a la faja y, por ende, a la polea ubicada en la parte superior, sobretodo; cabe precisar que el eje de esta polea superior es el que acciona el sistema de transmisión que pone a funcionar al generador.

El presente sistema está diseñado, de manera tal que, el agua no hace contacto ni con las poleas ni con la cadena, hecho que garantiza el periodo de vida útil de la cadena y de las poleas; además, el centro de gravedad del par de recipientes coincide tangencialmente con la circunferencia de la polea superior e inferior, lo que permite que el sistema funcione con mayor eficiencia.

Es pertinente aclarar que, según ciertos parámetros técnicos, es posible acoplar otra cadena paralela a la existente, este procedimiento es aconsejable cuando se disponga de un caudal que no puede ser aprovechado por los recipientes acoplados a una sola cadena, sobre la cual se encuentra distribuidos los pares de recipientes. El hecho de acoplar una segunda cadena permite aprovechar un caudal más elevado, por lo tanto, se obtiene un mayor peso, lo que significa una mayor potencia, ya que el peso del agua se acumula, ya no en dos recipientes, sino en tres recipientes. Al respecto, este tercer recipiente nos indica que la fuerza del peso del agua se distribuye entre las dos cadenas y en los tres recipientes, esta distribución de fuerzas es beneficioso para la durabilidad del sistema mecánico, y, a la vez, no afecta el rendimiento del sistema, pues los recipientes, en su conjunto, generan una sumatoria de fuerzas que hacen posible el giro de las poleas y, por ende, del generador también. Además, el uso de una segunda cadena y un tercer recipiente permiten acumular mayor cantidad de masa por cada unidad de distancia y por cada unidad de tiempo. En este caso los tres recipientes se alinean horizontalmente entre si y a su vez el centro de gravedad de cada recipiente se alinea horizontalmente con la cadena.

Descripción de las Figuras

FIGURA N° 1: Muestra una vista isométrica del sistema de recipientes para generación de energía eléctrica que comprende un sistema de recipientes (1), unos recipientes (1a y 1b), una cadena (2), una polea superior (3), un tubo (4), una varilla (5), un eje inferior (6), un tubo de alimentación (7), un eje superior (8), una polea inferior (9).

FIGURA N° 2: Muestra una vista frontal del sistema de recipientes para generación de energía eléctrica donde se muestra todos los elementos anteriormente mencionados.

FIGURA N° 3: Muestra una vista lateral derecha del sistema de recipientes para generación de energía eléctrica donde se muestra todos los elementos anteriormente mencionados.

FIGURA N° 4: Muestra una vista isométrica del sistema de recipientes para generación de energía eléctrica que comprende: un sistema de recipientes (1), unos recipientes (1a,1b y 1c), unas cadenas (2a y 2b) y todos los elementos anteriormente mencionados.

Realización Preferente de la Invención

Como se muestra en la figura 1, la máquina generadora de energía eléctrica presenta un sistema de recipientes (1) que comprende un recipiente (1a) unido mediante un tubo (4) ubicado en la parte inferior central del recipiente a otro recipiente (1b). Adicional a ello, una varilla (5) que une a ambos recipientes (1a y 1b) por la parte lateral central superior de ambos. La varilla (5) se encuentra acoplada a una cadena (2) mediante un tornillo de sujeción sobre la curva de un accesorio en forma de "U", donde cada uno de los extremos está soldado a un eslabón de la cadena (2), respectivamente siendo el punto de acople el centro de la varilla (5) y del tubo (4).

La cadena (2) une la polea superior (3) con la polea inferior (9) que se encuentran alineadas y separadas verticalmente, una respecto de la otra, y con sus ejes (6 y 8) en paralelo; los extremos de dichos ejes, descansan sobre sus respectivos rodamientos que están acoplados a una base que puede ser de madera y/o metal.

Las dos poleas (3 y 9), fijas sobre su respectivo eje, giran en un solo sentido, la polea superior (3) se fija a su respectivo eje superior (8) y la polea inferior (9) se fija a su respectivo eje inferior (6).

Como se muestra en las figuras, el invento funciona de la siguiente manera: una fuente alimentadora como un caudal, acorde con el diseño mecánico de la máquina, ingresa por un tubo de alimentación (7) que desemboca directamente sobre un recipiente (1b), el cual está conectado mediante un tubo (4) a otro recipiente (1a), produciéndose, de esta manera, un equilibrio del agua entre estos dos recipientes.

Según la altura y el caudal disponibles, cuando es necesario, un conjunto de recipientes interconectados se acopla sobre una cadena (2), de tal manera que, mediante el peso acumulado en los recipientes (1a y 1b), se transmite mayor potencia a la cadena (2) que, a su vez, transmite dicha potencia a la polea superior (3) y polea inferior (9) y, con ello, a sus ejes respectivos, eje superior (8) y eje inferior (6). En el caso de la polea superior (3), su eje superior (8) está conectado a un sistema mecánico amplificador de revoluciones que acciona a un generador de energía eléctrica. Es importante precisar que la potencia mecánica generada por la máquina está en concordancia con la potencia nominal del generador, según lo especifican los datos proporcionados por el fabricante.

La propulsión se inicia cuando el alimentador provee de una cantidad suficiente de agua a los dos recipientes (1a y 1b), enseguida, por efectos de la gravedad terrestre, éstos tienden a descender, generando, de esta manera, la propulsión. Cada par de recipientes (1) vierte el agua en el momento que gira sobre la polea inferior (9), momento a partir del cual los recipientes (1a y 1b) ascienden vacios para ser cargados otra vez, apenas giran sobre la polea superior (3), formándose, asi, un circuito constante. Cabe destacar que el volumen del agua es directamente proporcional a la potencia del sistema; la misma relación de proporcionalidad se establece respecto a la altura.

Como se muestra en la figura 4, el dispositivo generador de energia, también funciona sin inconvenientes acoplando otra cadena (2b) ubicada de forma paralela a la cadena (2a) precedente, por lo tanto, si bien la máquina puede funcionar al menos con una cadena con caudales y alturas inferiores a los de una turbina convencional, también puede funcionar con caudales iguales o superiores a las alturas y caudales de una micro central, por ejemplo; para lo cual, es suficiente con colocar otra cadena (2b), paralela y con la misma dimensión de la otra cadena (2a) sobre las cuales se acoplan sus respectivos recipientes (1a, 1c y 1b), estos recipientes tienen una ubicación paralela a las cadenas (2a y 2b); cada cadena une a dos poleas de la misma dimensión en cada extremo, estas poleas se encuentran fijadas sobre los ejes respectivos. Las dos poleas superiores giran sobre un eje común, lo mismo ocurre con las poleas inferiores.

El número de cadenas se puede aumentar de acuerdo a la cantidad de caudal disponible y acorde con la altura; por ejemplo según la figura 1, si se usa sólo una cadena (2), los recipiente son dos, tal como el primer recipiente (1a) y un segundo recipiente (1b), en este caso, el agua ingresa por el segundo recipiente (1b); pero como se muestra en la figura 4, si se utilizan dos cadenas (2a y 2b), los recipientes vendrian a ser tres: el primer recipiente (1a), el segundo recipiente (1b) y el tercer recipiente (1c), los tres alineados e interconectados por una varilla y un tubo, semejante a lo observado en la figura 1; en este caso el tercer recipiente (1c) va ubicado al centro de las dos cadenas y sobre este recipiente ingresa el agua, desde el cual se distribuye hacia los otros dos recipientes que se encuentran al lado derecho e izquierdo del tercer recipiente (1c); los recipientes (1a y 1b) se acoplan de tal manera que se ubiquen lateralmente, uno a cada lado de las cadenas (2a y 2b), siguiendo los nismos mecanismos de acoplamiento que se usa en el caso de una sola cadena, como se muestra en la figura 1.

Reivindicaciones

1. Un sistema de recipientes para generación de energía eléctrica que comprende al menos un par de recipientes (1), donde se une un primer recipiente (1a) con un segundo recipiente (1b), por la parte inferior de cada recipiente mediante un tubo (4), y por la parte superior lateral de cada recipiente mediante una varilla (5); la varilla (5) y el tubo (4) se encuentran acoplados a una cadena (2) en su parte central; la cadena (2) envuelve y une una polea superior (3) con una polea inferior (9), ambas poleas se encuentran ubicadas verticalmente una respecto de la otra; la polea superior (3) posee un eje superior (8) y la polea inferior (9), un eje inferior (6); la polea superior (3), mediante su eje superior (8), está conectado a un sistema mecánico amplificador de revoluciones que acciona a un generador de energía eléctrica.

2. Un sistema de recipientes para generación de energía eléctrica según reivindicación 1, caracterizado porque el primer recipiente (1a) y el segundo recipiente (1b) se sitúan en parte lateral de la cadena (2), y a una distancia estratégica de ésta, cuyo punto de acople con la cadena (2) es en la parte céntrica de la varilla (5) y del tubo (4).

3. Un sistema de recipientes para generación de energía eléctrica según reivindicación 1, caracterizado porque la cadena (2) mantiene una posición vertical que une a una polea superior (3) y a una polea inferior (9), éstas están alineadas y separadas en dirección vertical.

4. Un sistema de recipientes para generación de energía eléctrica según reivindicación 1, caracterizado porque la polea superior (3) y la polea inferior (9) están fijas sobre su respectivo eje que giran en un solo sentido, y el eje superior (8) así como el eje inferior (6) de la polea superior (3) y de la polea inferior (9), respectivamente, giran sobre sus respectivos cojinetes que se encuentran en los extremos de los mismos.

5. Un sistema de recipientes para generación de energía eléctrica según reivindicación 1, caracterizado porque presenta al menos dos cadenas (2a y 2b) sobre las cuales

se acopla unos recipientes (1a, 1b, 1c), siendo el ingreso del agua por el recipiente (1c) que está ubicado en la zona central respecto a los otros dos recipientes (1a y 1b) y también respecto a las dos cadenas (2a y 2b).

6. Un sistema de recipientes para generación de energía eléctrica según reivindicación 5, caracterizado porque cada una de las cadenas (2a y 2b) gira sobre dos poleas, una superior y otra inferior.

7. Un sistema de recipientes para generación de energía eléctrica según reivindicación 6, caracterizado porque las dos poleas superiores de ambas cadenas giran sobre un eje común, al igual que las dos poleas inferiores.

8. Un sistema de recipientes para generación de energía eléctrica según reivindicación 1, caracterizado porque el centro de gravedad de al menos un par de recipientes (1) coincide con los puntos tangenciales en los que concurren la cadena (2) y la polea superior (3) y, también la polea inferior (9).

Resumen

Se sabe que la tecnología que permite producir la hidroenergía, generalmente, lo constituyen los diversos tipos de turbinas; sin embargo, éstas operan a alturas y con caudales determinados, es decir, tienen un límite. Para superar estos inconvenientes, se han ideado otras formas de aprovechar la energía potencial del agua; entre estas formas tenemos las máquinas que permiten aprovechar energía potencial del agua mediante el uso de dos poleas unidas, ya sea por cables, fajas o cadenas sin fin, en los tres casos, con cubos o recipientes acoplados sobre las fajas, cables o cadenas. El inconveniente es que el periodo de durabilidad de los accesorios mecánicos mencionados, es corto, principalmente, porque el agua está en permanente contacto con estos elementos mecánicos, sobre todo con las poleas y las fajas, cables y cadenas, según sea el caso.

Ante esta situación se hizo una innovación tecnológica, que se refiere al uso de poleas con una cadena, sobre la cual van acoplados al menos un par de recipientes a cada lado de la cadena; es decir, un par cargado con agua y el otro, vacío. Cada recipiente se encuentra ubicado a una distancia de la parte lateral de la cadena, de tal manera que en el momento que se produce la carga y descarga del agua, ésta no hace contacto, ni con las poleas, ni con la cadena, propiciando, así, la posibilidad de lubricar la cadena y las poleas, lo que implica que los accesorios en referencia tengan un periodo de vida útil más prolongado.

En el caso de que se disponga de grandes caudales, es posible acoplar una segunda cadena, y con ello, se puede acoplar tres recipientes, este hecho permite acumular mayor cantidad de agua, por cada unidad de tiempo y de distancia, lo que implica el incremento de la potencia mecánica.

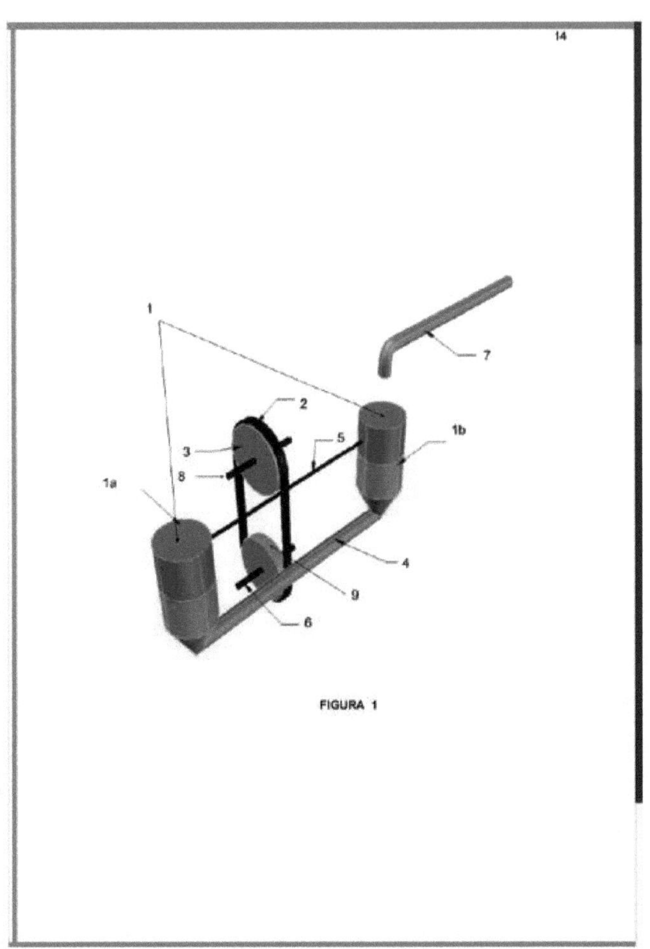

FIGURA 1

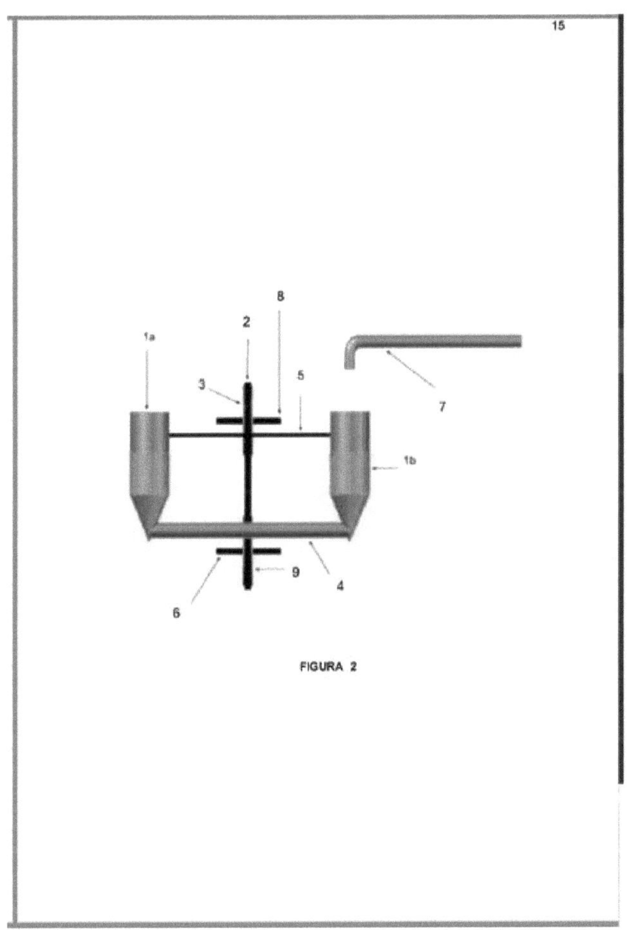

FIGURA 2

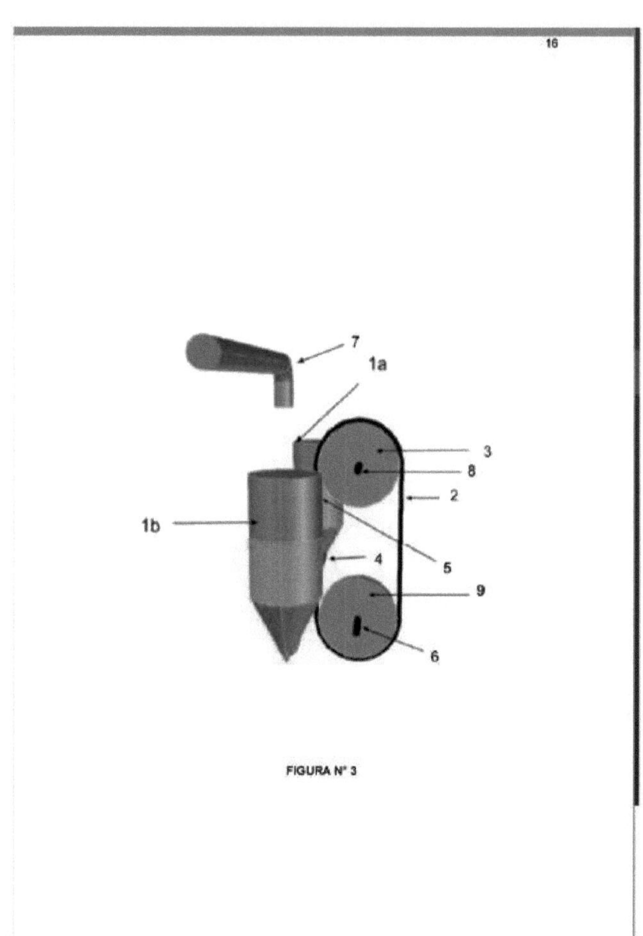

FIGURA N° 3

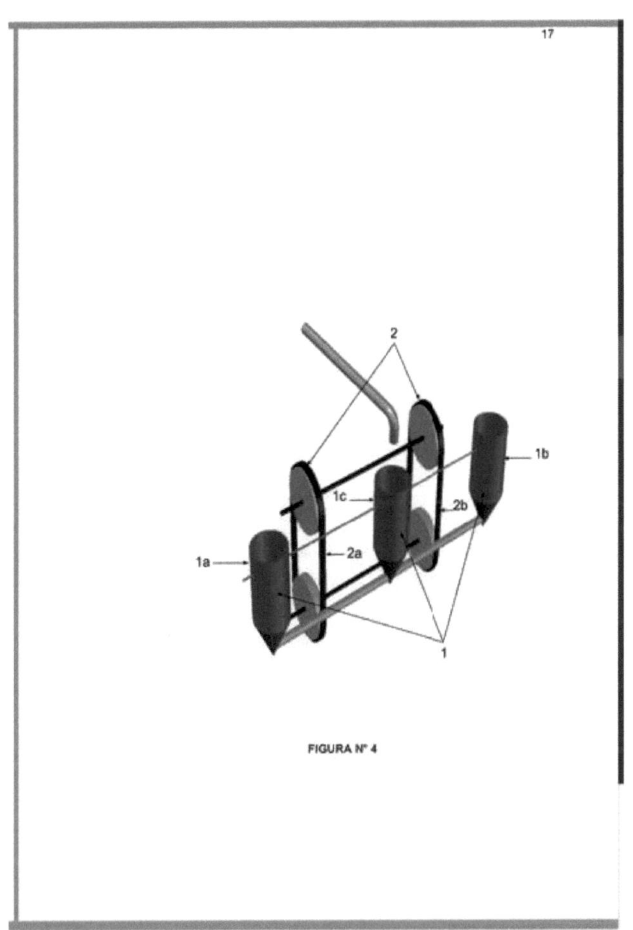

FIGURA N° 4

ANNEX N° 2

Universidad Nacional Mayor de San Marcos
Universidad del Perú. Decana de América

Vicerrectorado de Investigación y Posgrado

CESIÓN DE DERECHOS

Conste mediante la presente que el señor Agustin Esmaro Guevara Ruiz, identificado con DNI N° 40435591, inventor del **"Sistema de recipientes para generación de energía eléctrica"**, cede los derechos sobre dicha invención a la Universidad Nacional Mayor de San Marcos, identificada con RUC N° 20148092282.

Se cumple con legalizar la firma del inventor ante Notario Público.

Agustin Esmaro Guevara Ruiz
DNI N° 40435591

CERTIFICO: la autenticidad de la firma que antecede de: Agustin Esmaro Guevara Ruiz con DNI No. 40435591 que legalizo

Callao, 12 OCT. 2018

Manuel Galvez Succar
ABOGADO NOTARIO

EL NOTARIO NO ASUME RESPONSABILIDAD SOBRE EL CONTENIDO DEL DOCUMENTO (ART. 108 DEL D.L. 1049)

DOCUMENTO NO REDACTADO EN ESTA NOTARIA

Av. Germán Amézaga s/n – Ciudad Universitaria
Biblioteca Central UNMSM, 4° Piso

Central Telefónica 619-7000, Fax 7377

ANNEX N° 3

SOLICITUD DE REGISTRO DE PATENTE

Indecopi

(21) 144209

2018 OCT 17 PM 4 54

RECIBIDO

DIRECCIÓN DE INVENCIONES Y NUEVAS TECNOLOGÍAS

(22) 002013-2018/DIN

A la Dirección de Invenciones y Nuevas Tecnologías se solicita el registro de la concesión, conforme a las siguientes especificaciones, de:

(12) Patente de Invención ☐ Modelo de Utilidad ☒

(71) Solicitante (s), domicilio (s), y país (es)

UNIVERSIDAD NACIONAL MAYOR DE SAN MARCOS
Calle Germán Amézaga N°375 – Edificio Jorge Basadre, Ciudad Universitaria, Lima 1

Teléfono (s) 619 7000 anexo 7577 Telefacsímil (es)

(72) Inventor (es), domicilio (s) y nacionalidad (es)

AGUSTÍN ESMARO GUEVARA RUÍZ
Jr. Diana Mz. B Lt. 8, Surco, Lima
Peruano

(74) Representante / Apoderado y domicilio

FELIPE ANTONIO SAN MARTÍN HOWARD
Calle Germán Amézaga N°375 – Edificio Jorge Basadre, Ciudad Universitaria, Lima 1

N° Agente Poder N° Anexo a:

Teléfono (s) 619 7000 anexo 7577 Telefacsímil (es)

(54) Título de la invención

SISTEMA DE RECIPIENTES PARA GENERACION DE ENERGIA ELECTRICA

(51) Clasificación internacional sugerida (CIP[7]) F03B9/00

(30) Reivindica prioridad Si ☐ No ☒

(31) Número (s) (32) Fecha (s) (33) País (es)

INSTITUTO NACIONAL DE DEFENSA DE LA COMPETENCIA Y DE LA PROTECCIÓN DE LA PROPIEDAD INTELECTUAL
DIRECCIÓN DE INVENCIONES Y NUEVAS TECNOLOGÍAS
Calle De la Prosa 104, San Borja, Lima 41 - Perú Telf: 224-7800 Anexos 3805, 3806, 3801 o 3811
Web: www.indecopi.gob.pe

F-DIN-01/01

DECLARACIÓN SOBRE UTILIZACIÓN DE RECURSO GENÉTICOS Y/O CONOCIMIENTOS TRADICIONALES:

1. Declaro que mi invención fue obtenida o desarrollada a partir de recursos genéticos o de sus productos derivados de Países Miembros de la Comunidad Andina.

☐ SI. Indique el lugar de colecta o extracción ____________________

☑ NO

2. Declaro que mi invención fue obtenida o desarrollada a partir de conocimientos tradicionales de las comunidades indígenas, afroamericanas o locales de Países Miembros de la Comunidad Andina.

☐ SI. Indique el lugar de colecta o extracción ____________________

☑ NO

RECAUDOS ANEXOS:	**Fecha: 15 de octubre de 2018**
☒ Descripción: ____ hojas (____ ejemplares) ☒ Reivindicaciones: ____ hojas (____ ejemplares) ☒ Resumen (____ ejemplares) ☒ Dibujos o Planos: - numerados de Nº ____ a Nº ____ (____ ejemplares) - se sugiere dibujo Nº ___ para la publicación ☒ Poder o documento de personería ☐ Documento(s) de prioridad ☐ Certificado de exhibición ☒ Comprobante de pago de tasa ☐ Informe(s) de búsqueda o de patentabilidad extranjero(s) ☐ Reducciones del plano o dibujo principal ☒ Documento de cesión ☒ Otros, especificar: PATENTA – MODALIDAD CENTROS ACADÉMICOS Y DE INVESTIGACIÓN	**Firma** **Felipe Antonio San Martín Howard**

En cumplimiento de lo dispuesto por la Ley Nº 29733, Ley de Protección de Datos Personales, le informamos que los datos personales que usted nos proporcione serán utilizados y/o tratados por el Indecopi (por sí mismo o a través de terceros), estricta y únicamente para administrar el sistema de promoción, registro y protección de derechos de propiedad intelectual (signos distintivos, invenciones y nuevas tecnologías, y derecho de autor) en sede administrativa, así como, de ser el caso, para las actividades vinculadas con el registro de usuarios del sistema de patentes, pudiendo ser incorporados en un banco de datos personales de titularidad del Indecopi.
Se informa que el Indecopi podría compartir y/o usar y/o almacenar y/o transferir su información a terceras personas, estrictamente con el objetivo de realizar las actividades antes mencionadas.
Usted podrá ejercer, cuando corresponda, sus derechos de información, acceso, rectificación, cancelación y oposición de sus datos personales en cualquier momento, a través de las mesas de partes de las oficinas del Indecopi.

ANNEX N° 4

Feria de Inventos y Diseños Industriales

EXPO PATENTA

LA CREATIVIDAD DE LOS INVENTORES PERUANOS

Patenta

Se otorga el presente Diploma a:

AGUSTÍN ESMARO GUEVARA RUÍZ

Por el invento denominado

SISTEMA DE RECIPIENTES PARA GENERACIÓN DE ENERGÍA ELÉCTRICA

el cual participó en el XVII Concurso Nacional de Invenciones y Diseños Industriales realizado del 22 al 25 de noviembre del 2018 por el Instituto Nacional de Defensa de la Competencia y de la Protección de la Propiedad Intelectual – INDECOPI.

Manuel Castro Calderón
Director
Dirección de Invenciones y Nuevas Tecnologías
INDECOPI

ANNEX N° 5

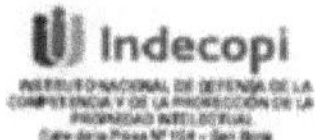

DIRECCION DE INVENCIONES Y NUEVAS TECNOLOGÍAS

Examen de patentabilidad	
JDB	008-2020

Expediente:	002013-2018/DIN	Fecha de ingreso:	2018-10-17
		Fecha de presentación:	2018-10-17

Solicitud internacional PCT	N°	
	Fecha:	

Solicitante(s):	UNIVERSIDAD NACIONAL MAYOR DE SAN MARCOS

Prioridad(es)	Fecha(s) de prioridad(es)
No reivindica	

Modalidad de protección	Patente de Invención (Decisión 486 / Art. 45)	
	Modelo de Utilidad (Decisión 486 / Art. 45 concordante con Art. 85)	X
Título:	«Sistema de recipientes para generación de energía eléctrica»	

Referencia(s):	

Opositor(es):	

1. Textos a analizar	Originalmente presentado	Modificaciones	Fecha	Ampliación[1] (Decisión 486 / Art. 34)	
				SI	NO
Memoria descriptiva	X				
Figura(s)	X				
Reivindicación(es)	1 a 8				
Resumen	X				
Listado de secuencias					
Otros[2]					

[1] En caso de que las modificaciones realizadas impliquen una ampliación de lo originalmente presentado, esta nueva documentación modificada no será tomada en cuenta y toda opinión final emitida se realizará en función a la documentación original.

* Constituye una ampliación parcial.

[2] e.i. argumentos, resultados de ensayos, cuadros comparativos, etc.

1/4

Oposición					
Respuesta a oposición					

	Sí	No
2. No invenciones y excepciones a la patentabilidad (Decisión 486 / Art. 15, 20 y/o 82[3])		1 a 8
3. Usos (Decisión 486 / Art. 14[4] y 21)		1 a 8
4. Unidad de invención[5] (Decisión 486 / Art. 25)	1 a 8	

5. Requisitos a evaluar	Cumple	No cumple
5.1 Para solicitud fraccionaria		
No amplía divulgación de la solicitud inicial (Decisión 486/art. 36)		
Reivindicaciones: No genera doble protección de la solicitud inicial (D.L. 1075/art. 29)		
5.2 Suficiencia, claridad, concisión y soporte		
Memoria descriptiva (Decisión 486/art. 28)	X	
Reivindicaciones (Decisión 486/art. 30)	1 a 8	
5.3 Novedad (Decisión 486/art. 16)		
5.4 Nivel inventivo (Decisión 486/art. 18)		
5.5 Aplicación industrial (Decisión 486/art. 19)		
5.6 Para modelos de utilidad (Decisión 486/art. 81[3])	1 a 8	

[3] En los casos de patentes de Modelo de Utilidad.
[4] Interpretación por el Tribunal de Justicia de la Comunidad Andina que, a la luz del Proceso N° 89-A-2000 considera patentables los productos o los procedimientos, mas no los usos.
[5] En caso de que se encuentre más de un concepto inventivo en la presente solicitud y se identifique más de un grupo de reivindicaciones relacionadas con dichos conceptos inventivos, se procederá con el análisis respecto al primer grupo encontrado.

6	**Análisis y opinión escrita acerca de los requerimientos considerados en los numerales anteriores**

6.1 Cita(s) y documentación consideradas para la emisión del examen:

Antecedente(s) relevante(s) del estado de la técnica:

D1= US 2005/0052028 publicado el 2005-03-10
(Chiang Kud-Chu)
«Sistema de generación de energía hidráulica basado en bombeo por peso de agua»

6.2 Formulación de opinión y argumentos respecto al (a los) punto(s):

Respecto del numeral 5.6 (Para modelos de utilidad)

Novedad y ventaja técnica

La **reivindicación 1** referida a un sistema de recipientes para generación de energía eléctrica se diferencia del documento D1 en que comprende:

- al menos un par de recipientes (1), donde se une un primer recipiente (1a) con un segundo recipiente (1b) por la parte inferior de cada recipiente mediante un tubo (4), y por la parte superior lateral de cada recipiente mediante una varilla (5), y donde la varilla (5) y el tubo (4) se encuentran acoplados a una cadena (2) en su parte central;

mientras que el sistema de generación de energía hidráulica descrito en D1, comprende unas cubetas de accionamiento (11-23) individuales unidas a un eje de transmisión (25) por medio de solo unos ejes de cubeta (19) individuales **(ver *resumen, párrafos [0018] y [0020] y figura 1 de D1*)**; por lo tanto, la **reivindicación 1 tiene novedad.**

Comparación gráfica

Reivindicación 1	Documento D1

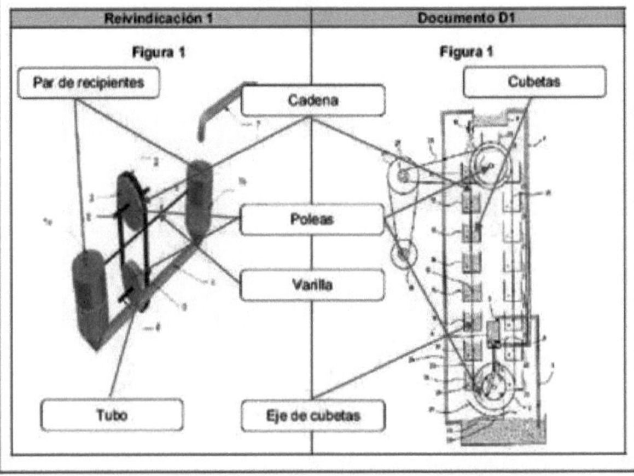

Ventaja técnica

La varilla (5) y el tubo (4) permiten que al menos un par de recipientes (1) del sistema de recipientes para generación de energía eléctrica de la invención sean dispuestos lateralmente en relación a la cadena (2), evitando que el agua no haga contacto con la cadena (2) ni con las poleas (3, 9), tal como se menciona en el penúltimo párrafo de la tercera página y el tercer párrafo de la cuarta página de la memoria descriptiva; por lo que, la **reivindicación 1 tiene ventaja técnica.**

Por lo tanto, la **reivindicación 1 cumple** con los requisitos establecidos en el **artículo 81** de la Decisión 486 de la Comunidad Andina.

Las **reivindicaciones 2 a 8**, al ser dependientes de la reivindicación 1, **también cumplen** con los requisitos establecidos en el **artículo 81** de la Decisión 486 de la Comunidad Andina.

7.Conclusión(es)	**Fecha de emisión: 2020-03-26**
En base a la Decisión 486 de la Comisión de la Comunidad Andina: Las reivindicaciones 1 a 8 cumplen con los requisitos establecidos en el artículo 81 de la Decisión 486 de la Comunidad Andina.	**Elaborado por:** Jesús Grabiel Diestra Balta Examinador de Patentes CIP 175429

Revisado por:

NELSON ALEXANDER CRUZ TAPIA
Especialista 2
Dirección de Invenciones y
Nuevas Tecnologías
INDECOPI

ANNEX N° 6

DIRECCIÓN DE INVENCIONES
Y NUEVAS TECNOLOGÍAS

Reporte de búsqueda JDB 008-2020	**Expediente n.°** 002013-2018/DIN
Fecha de ingreso: 2018-10-17	**Fecha de presentación:** 2018-10-17
C.I.P. (8) : F03B 1/02; F03B 7/00	**Fecha de prioridad:** No reivindica

Términos de búsqueda empleados:

Sistema; recipientes; energía; hidráulica; laterales; eje; cadenas; agua.

F03B 1/02; F03B 7/00

Documentos considerados relevantes

Categoría	Cita del documento, indicando las partes pertinentes y la fecha de publicación	Reivindicaciones afectadas
A	US 2005/0052028 publicada el 2005-03-10 (Chiang Kud-Chu) Ver resumen, párrafos [0018] y [0020] y figura 1. https://worldwide.espacenet.com/patent/search/family/034076560/publication/US2005052028A1?q=pn%3DUS2005052028A1	1 a 8

Categoría de documentos citados:

X: Particularmente relevante por sí solo.
A: Estado de la técnica general, no particularmente relevante.
O: Divulgación oral.
E: Solicitud presentada antes pero publicada después de la fecha de presentación de la solicitud examinada (sólo con X).
D: Citado en la solicitud.
L: Citado por otras razones.
P: Anterior a la fecha de presentación pero posterior a la fecha de prioridad.
T: Teoría o principio en el que se basa la invención
&: Documento miembro de la misma familia de patentes.

Examinador: **Ing. Jesús Grabiel Diestra Balta**

ANNEX N° 7

DIRECCIÓN DE INVENCIONES Y NUEVAS TECNOLOGÍAS

EXPEDIENTE N° 002013-2018/DIN

RESOLUCIÓN N° 001313-2020/DIN-INDECOPI

Lima, 05 de noviembre de 2020

Patente de modelo de utilidad: Concedida

Mediante expediente N° 002013-2018/DIN, iniciado el 17 de octubre de 2018, UNIVERSIDAD NACIONAL MAYOR DE SAN MARCOS de Perú, solicita patente de modelo de utilidad para "SISTEMA DE RECIPIENTES PARA GENERACIÓN DE ENERGÍA ELÉCTRICA", C.I.P.8 F03B 1/02; F03B 7/00, cuyo inventor es Agustín Esmaro GUEVARA RUIZ.

1. **EXAMEN DE PATENTABILIDAD**

El modelo de utilidad solicitado reúne los requisitos establecidos en la Decisión 486 de la Comisión de la Comunidad Andina que aprueba el Régimen Común sobre Propiedad Industrial, conforme aparece en el examen de patentabilidad que corre de fojas 37 a 39 del expediente.

La presente Resolución se emite en aplicación de la norma legal antes mencionada y en uso de las facultades conferidas por los artículos 37 y 40 de la Ley de Organización y Funciones del Instituto Nacional de Defensa de la Competencia y de la Protección de la Propiedad Intelectual (Indecopi) sancionada por Decreto Legislativo N° 1033, concordado con el artículo 4 del Decreto Legislativo 1075 que aprueba las disposiciones complementarias a la Decisión 486 de la Comisión de la Comunidad Andina.

2. **RESOLUCIÓN DE LA DIRECCIÓN DE INVENCIONES Y NUEVAS TECNOLOGÍAS**

OTORGAR patente de modelo de utilidad para "SISTEMA DE RECIPIENTES PARA GENERACIÓN DE ENERGÍA ELÉCTRICA", C.I.P.8 F03B 1/02; F03B 7/00, a favor de UNIVERSIDAD NACIONAL MAYOR DE SAN MARCOS de Perú, por un plazo de diez (10) años, contados desde el 17 de octubre de 2018, fecha de presentación de la solicitud, aprobándose las 8 reivindicaciones que corren a fojas 13 y 14 del expediente.

Regístrese y Comuníquese

MANUEL CASTRO CALDERÓN
Director de Invenciones y
Nuevas Tecnologías
INDECOPI

JMS/jsa

INSTITUTO NACIONAL DE DEFENSA DE LA COMPETENCIA Y DE LA PROTECCIÓN DE LA PROPIEDAD INTELECTUAL
Calle De la Prosa 104, San Borja, Lima 41 - Perú Telf: 224 7800 / Fax: 224 0348
E-mail: postmaster@indecopi.gob.pe / Web: www.indecopi.gob.pe

M-DIN-15/01 1 de 1

ANNEX N° 8

INFORME

A : UNIVERSIDAD NACIONAL MAYOR DE SAN MARCOS.

DE : JOSE LUIS FABIAN CHALE.
Consultor en Propiedad Industrial.

ASUNTO : Consultoría preliminar del Proyecto: "Sistema mecánico para generación de energía eléctrica a pequeña escala por propulsión gravitatoria"; presentado por Agustín Esmaro Guevara Ruiz.

---

1. **DESCRIPCIÓN.-**

El proyecto se refiere a un sistema mecánico para generación de energía eléctrica a pequeña escala por propulsión gravitatoria que comprende un alimentador de agua dispuesto en la parte superior, dos poleas alineadas en dirección vertical, una superior y otra inferior, entre las cuales se coloca una cuerda o cable; sobre la cuerda se acoplan cubos o baldes dispuestos separadamente en forma uniforme con sus bocas orientadas en una misma dirección; sobre un lado los baldes están con la boca hacia arriba y en el otro lado se encuentran con la boca hacia abajo; durante su funcionamiento el balde dispuesto en la parte superior con la boca arriba es llenado con agua y debido al peso del agua el balde se desplaza hacia la parte inferior donde al cambiar de dirección descarga el agua y luego es elevado hacia la parte superior; un dinamo o generador se acopla a una de las poleas para generar energía eléctrica o para cargar baterías.

2. **ANTECEDENTES ENCONTRADOS:**

El antecedente más cercano encontrado es el siguiente:

D1 = Publicación del VIII Concurso de Inventores nacionales, INDECOPI 2004-11-04; Mini hidroeléctrica, presentado por Agustín Esmaro Guevara Ruiz (Ver anexo adjunto).

3. **NOVEDAD Y PATENTABILIDAD**

El proyecto propuesto es anticipado por el documento D1. Por lo tanto, NO sería factible su patentamiento.

JOSE LUIS FABIAN CHALE.
Consultor en Propiedad Industrial.

1

ANNEX N° 9

GLOSSARY

INDECOPI: The National Institute for the Defence of Competition and Intellectual Property Protection (INDECOPI) is a specialised public body attached to the Presidency of the Council of Ministers. It started its activities in November 1992, by means of Decree Law N°25868. Its functions are the promotion of the market and the protection of consumer rights. It also promotes a culture of fair and free competition in the Peruvian economy, safeguarding all forms of intellectual property: from distinctive signs and copyrights to patents and biotechnology. As a result of its work in the promotion of fair and free competition rules among the agents of the Peruvian economy, INDECOPI is conceived as a service entity that promotes the development of a culture of quality, to achieve the full satisfaction of citizens, entrepreneurs and the State (Indecopi, 2020, n. p.).

Invention: Any new technical solution to a technical problem in any field of technology *(*fUNMSM, 2018, p.04).

Inventor: A natural person who has generated a useful and novel creation of industrial application (UNMSM, 2018, p.04).

Utility model patent: It is any new form, configuration or arrangement of elements, of any device, tool, mechanism or other object, or of any part thereof, that allows a better or different operation, use or manufacture of the object that incorporates it, or provides it with any utility, advantage or technical effect that it did not have before (UNMSM, 2018, p.05).

Technology: Is any system of founded practical techniques, or study of them, distinguishing it from the simple technique or the precientific technique (Bunge, 2012, p. 51).

Technologist: The technologist applies the scientific method to problems of practical interest (Bunge, 2012, p. 51*)*.

MINUTES OF SUPPORT

Escuela Académico Profesional de Filosofía

ACTA DE SUSTENTACIÓN DE TESIS

PARA OBTENER EL TÍTULO PROFESIONAL DE LICENCIADO EN FILOSOFÍA

Reunido el Jurado en sesión virtual, el día miércoles 25 de noviembre de 2020 a las diez horas, integrado por el Mg. Dante Dávila Morey (Presidente), Dr. Luis Adolfo Piscoya Hermoza (Asesor), Lic. Aníbal Campos Rodrigo (Informante) y Mg. Herminio Paucar Curasma (Informante) para calificar la sustentación de la tesis titulada **LA TELEOLOGÍA DE LOS EXPERIMENTOS CIENTÍFICOS: EL CASO DE LA CAÍDA LIBRE DE LOS CUERPOS**, presentada por el bachiller Agustín Esmaro Guevara Ruiz, para optar el título de Licenciado en Filosofía.

Después de la exposición del tesista, la lectura de sus conclusiones y absueltas las preguntas formuladas por el Jurado, este se retiró a deliberar y acordó la siguiente calificación de acuerdo a lo establecido por el Reglamento General de Estudios de Pregrado:

Sobresaliente con mención (20)

Habiendo sido aprobada la sustentación de la tesis, el Jurado recomendó que la Facultad proponga que se le otorgue el título de Licenciado en Filosofía al bachiller Agustín Esmaro Guevara Ruiz.

Concluido el acto académico a las 12:30 horas, firman la presente acta.

Mg. Dante Dávila Morey
Presidente

Lic. Aníbal Campos Rodrigo
Jurado Informante

Mg. Herminio Paucar Curasma
Jurado Informante

Dr. Luis Adolfo Piscoya Hermoza
Jurado Asesor

Letras mayúsculas del Perú y América
Facultad de Letras y Ciencias Humanas / Universidad Nacional Mayor de San Marcos
Calle Germán Amézaga n.° 375, Lima 1 - Perú. Ciudad universitaria (puerta 3)
Teléfonos: (051) (01) 452 4641 / (051) (01) 619 7000 - www.letras.unmsm.edu.pe

Printed by Books on Demand GmbH, Norderstedt / Germany